How To Be
Your Own
HOME
ELECTRICIAN

*the text of this book is printed
on 100% recycled paper*

How To Be
Your Own
HOME
ELECTRICIAN

By
GEORGE DANIELS

BARNES & NOBLE BOOKS

A DIVISION OF HARPER & ROW, PUBLISHERS

New York, Hagerstown, San Francisco, London

CONTENTS

	Introduction	7
	A Few Words about the National Electrical Code	14
1.	Tools for Wiring	16
2.	Types of Wires and Wiring Techniques	21
3.	How to Install House Wiring	30
4.	Wiring for Heavy Loads	50
5.	Wiring an Old House	54
6.	Surface Wiring	57
7.	Outlet Repairs and Installations	68
8.	Plugs and Cords	78
9.	Wall and Ceiling Fixtures	87
10.	Installing and Replacing Switches	95
11.	Outdoor Wiring	107
12.	Testing Wiring	116
13.	Fluorescent Lights	118
14.	Doorbells and Chimes	123
15.	Fuses	127
16.	Working with Electric Motors	132
	Index	141

INTRODUCTION

ELECTRICITY AND ITS MEASUREMENT. Theory tells us that an electric current traveling through a wire is a movement of electrons—about 6.28 billion billion for each ampere. So, if you're reading this by the light of a lamp with a pair of 60-watt bulbs, you have about 6,280,000,000,000,000,000 electrons hustling through the lamp cord every second. You can't see them or weigh them as they do their work, but you can measure the current they produce. And you measure it much as you would measure water flowing through a pipe. Instead of figuring in gallons per minute, however, as with water, you calculate electricity in "coulombs" per second. But you're not likely to hear that term often, as a current of 1 coulomb per second is called a current of 1 *ampere*. That is the more convenient term you see abbreviated to "amps." on the specification plates of the electric motors you use.

The pressure or "push" that moves the piped water along is measured in pounds per square inch. Similarly, the push behind an electric current is measured in volts. And here again, terms may be combined for convenience. Multiply the number of amps. a device consumes by the number of volts in the power line and you have its rating in *watts*—the measuring units you see marked on light bulbs, toasters, and electric heaters. (On alternating current, this simple mathematics doesn't apply to such things as motors and buzzers because of technical factors.)

WHY POWER COMPANIES SUPPLY ALTERNATING CURRENT. The type of current supplied to most American homes today is 60-cycle alternating current, commonly called AC. Unlike direct current (DC), as from a battery, which always flows in the same direction, AC reverses its direction of flow sixty times a second at 60 cycles.

Power companies use AC because the rapid reversal creates electrical effects that enable them to do essential things that can't be done with DC. For example, they can use a *transformer* (which won't work on ordinary DC) to "step up" voltage for long-distance transmission, while automatically lowering

7

amperage proportionately. (Transmission losses are much greater with high amperage and low voltage than with high voltage and low amperage.) At the destination point another transformer is used to "step down" the voltage and automatically raise the amperage to provide a usable combination. There's a transformer on a utility pole not too far from your house to reduce the several thousand volts carried by the main power lines to the 120 volts required by your lamps and appliances. In most modern systems it also provides the 240 volts needed by heavier equipment.

In greatly simplified terms the transformer consists of two completely separate coils of wire wrapped around the same soft-iron core. When AC is fed into the primary (incoming) coil, the rapid magnetic changes in the iron core "induce" a separate AC in the *secondary* (outgoing) coil. And the voltage in the separate coils is proportionate to their relative sizes. If the secondary coil has ten times the number of turns as the primary, it will have ten times the voltage, and you have a "step-up" transformer. When the sizes are the other way around, the outgoing current is lower and you have a "step-down" transformer. This is the same principle used in the transformers that operate a toy train set.

The AC is produced at the power plant by *alternators*, which are giant versions of those used in many cars. Simple devices are often used to explain their operation and demonstrate the principle involved. If, for example, you move a wire downward between the poles (tips) of a horseshoe magnet, it cuts across invisible magnetic lines of force between the magnet tips, and an electric current is actually generated in the wire. If you then move the wire upward along the same path, a new current is generated in it, but the new current is traveling in the opposite direction. An alternator is simply a machine that does the same job (with rotary motion) on a much larger scale.

The relationship between magnetism and electric current plays a part in many of the things electricity does for you though you are seldom aware of it. An electric motor, for example, gets its power simply by harnessing magnetism so its attraction or repulsion will spin an armature and, in turn, the motor shaft. *Electromagnets* make it possible. Made in many shapes to fit their use, they are essentially similar—a soft-iron core wrapped with a coil of wire. Pass a current through the wire and the core becomes magnetized. Shut off the current and most of the magnetism vanishes. Contacts in electric motors switch electromagnets on and off at precisely the right points to utilize their attraction to spin the rotor and shaft.

You can easily make a small electromagnet by wrapping a large nail or bolt with fine wire (one type is called magnet wire) and attaching the ends to a large dry cell. With the wire attached to the battery, the nail is magnetized, and can be a handy tool for picking up spilled tacks or brads. Place the miniature electromagnet, with its cluster of tacks or brads, over a box, disconnect one end of the wire, and the cluster will drop into the box as the magnetic attraction terminates.

So much for the theoretical aspects of electricity. It would require many

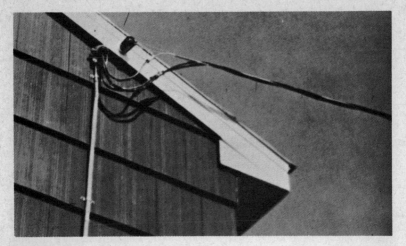

Service head starts wires high on house to clear objects in yard, roadway, etc. Cable runs down to meter socket in this installation.

books to cover the subject thoroughly. But, with some of the basic facts in mind, the practical work of house wiring may seem more interesting; and the reasons for certain methods and precautions will be more understandable.

WHERE YOUR WIRING WORK BEGINS. House wiring starts at the *service entrance,* the point where the power comes into your house. The utility company's wires are led in through the *service head* mounted high on your house to keep the wires well above the ground. From the service head they run down the outside of the house to the meter socket that contains the meter on completed jobs. From there they are led into the house (through a fitting made for the purpose) to the *service panel.* This contains the fuses or circuit breakers that protect the inside wiring from short circuits and overloads. It also provides one of several forms of main switch for shutting off all current to the inside wiring, so you can make repairs or modifications safely with all wiring dead.

Complete details of the service-entrance installation and the wiring emanating from it are given in a later chapter. But the brief summary of these wiring stages in the following paragraphs will enable you to follow the detailed instructions much more rapidly.

FEATURES THAT MAKE WIRING EASY AND SAFE. On the back of the hinged cover of most service panels, you will find a diagram showing exactly how to connect the wiring for each house circuit to the terminal screws provided in the panel. From that point on through the house every wire and every connecting screw is color coded, or in some instances, labeled. Whether you are using cable, conduit, or raceway, you will be installing just two wires, a black one and a white one, most of the time. And the working rules are so simple that errors in connection are next to impossible. *You need only remember that in all joints and splices the white wire always connects to a white wire, black always to black.* Never connect one color to another. (There is only one exception to this rule in a specific switching arrangement described and illustrated in the chapter dealing with switches.)

When you install outlet receptacles you will notice that they have one pair of chrome connecting screws and one pair of brass connecting screws. *The white wire is always connected to one of the chrome screws, the black wire always to one of the brass screws.* The remaining chrome and brass screws are used to attach the continuing run of wiring (in the same manner) if more outlets are to be installed beyond the first one.

Switches, you will note, have brass screws only, and there's an important reason. *Only the black wire is ever connected to a switch, never the white wire.* The reason: the white wire is always the *ground* wire, and must have a continuous unbroken run through all connections back to the service panel. At that point the grounded or neutral part of the circuit is actually attached to the earth by a wire leading to the water pipes of the house or a metal rod driven in the ground. (This grounding is very important. If lightning should strike the wires outside the house, it has a direct path to the earth, greatly reducing the chance of scattering its effects through the house with possibly fatal results.)

Fixtures for wall and ceiling lights are chrome and brass coded at their connections the same as outlet receptacles. The screws that hold these electrical items in the metal boxes that house them are standardized, as is their spacing. The fittings that lock each type of cable into these boxes are also standardized. All you need do is make certain that you buy the type that matches the cable or wiring system you are using. If it happens to be raceway, you will find fittings available to connect it to any other form of approved wiring.

House wiring is really very similar to assembling a custom-tailored do-it-yourself kit. Buy only materials and components that carry the Underwriters' Laboratories label, as this assures you that the items meet at least minimum approved safety standards. Underwriters' Laboratories, Inc. is a group of electrical labs which test and approve products submitted to them by manufacturers. If the product meets Underwriters' specifications, it is labeled as such. If you follow the basic assembly rules in the chapters that follow, your wiring will be safe and the job will be easy.

ABOUT SAFETY AND CODES. In all electrical work keep safety foremost in your mind. Work on any wiring only when the current is definitely *off*. And, in making connections to a service panel that is already in use, remember that there are live connections on the power-line side of the switch even when the switch is off. Keep well clear of them. And do not be misled by terminology. The black and red wires of a 120–240 volt wiring-system, for example, are often referred to as the "hot" wires. The white wire is commonly called the ground or "neutral" wire. Never acquire the misconception that the white wire is any less dangerous when the current is turned on simply because it is not called a "hot" wire. It is one side of the circuit, and from the safety standpoint is just as "hot" as either of the others.

Before beginning any house-wiring work, find out if there is a local code in your area, if you require a permit before doing the work, and if inspections will be made. The inspections are often made before the wiring is entirely completed so that internal details are visible. Check the details of the local code, of course, in advance. It may differ from the national code. Your work must comply with it in order to avoid violations that can take considerable time to remedy, increasing your costs and delaying the job. In general, however, the National Electrical Code is the accepted guide to safe materials and wiring practices. It does not tell you how to do a wiring job, but it clearly defines practices, and designates materials, that have proven safe. Study it before beginning your wiring.

UTILITY COMPANIES. If you are planning a complete house wiring job, as in the course of building your own home, consult your local power company *before* you begin. Various factors can determine the part of your house at which the service entrance must be installed. Because of the location of utility poles on the opposite side of the street, for example, it may be necessary to run your entrance wires to one particular end of the house to avoid having them traverse someone else's property. Or it may be necessary to run them to a particular corner of the house to prevent them from passing too low over a highway. If your house is a modern one-story design, it may also be neces-

Meter socket before meter is installed by power company (left). Lower cable runs to entrance panel. Smaller cable (resembling BX) is ground wire, leads to connection with water pipe. When wiring is complete, meter is installed and power connected (right).

Inside house, entrance panel and main switch unit distributes current to individual circuits. This one is open, will have cover and door when completely wired.

sary to install an entrance mast to hold the wires high enough above the ground. As the location of the service entrance determines the general layout of your wiring (as all interior wiring ultimately leads to that point), much extra work could be entailed in revision to suit an unexpected entrance location.

The power company can also give you complete information on the amount of work the company's men will do in connection with the service entrance. This varies. In a sample instance the homeowner does all wiring to the entrance panel and provides the materials from there to the meter socket, and up to the entrance head. Then the power company connects its wires, installs the meter, and the system is ready for use. In most cases the company sends a man in advance to check over the location of your house and any factors that may affect exterior wiring. As a rule he will also be able to supply you with valuable information regarding local codes and regulations.

Usually, too, you will have a choice of service-entrance capacities, often ranging from 60 amp. service (the minimum to "get by" according to the National Electrical Code) up to 200 amp. service to provide enough reserve for just about any appliances and power tools you are likely to acquire in the future. A service of 100 amps. is recommended by the Adequate Wiring Bureau for homes up to 3000 square feet in floor area. Your best bet: select as large a capacity service as your budget permits. (The larger service panels cost more.) This way you are less likely to be limited later on in your choice of large appliances.

PLANNING YOUR WIRING. If you are wondering how many outlets your various rooms should have you may find the National Code very helpful as a general guide. If a wiring job is to comply with the code (it does not have the force of law) there must be enough *lighting* circuits to supply 3 watts for every square foot of floor space in the house. How far apart should you space your outlets? Again, the code offers a formula: "Receptacle outlets shall be installed so that no point along the floor line in any usable wall space is more than 6 ft., measured horizontally, from an outlet. . . ." Practically, this means keep your outlets no farther than 12′ apart when you plan your layout.

Notice that the circuits mentioned in the preceding paragraph are *lighting* circuits. These are intended to supply your lighting fixtures and power such appliances as vacuum cleaners, radios, and TV sets. You'll need *special appliance circuits*, too, for heavier-load items like toasters and electric irons. And for major appliances like electric ranges and automatic laundries, you'll need *individual appliance circuits*. The greater the power required by an appliance, the greater the capacity of the wires must be, as explained in a later chapter. But the wiring, itself, does not become more complicated. In general, lighting circuits are fused for 15 amps and may be carried by No. 14 wire. Small-appliance circuits are fused for 20 amps and require No. 12 wire, although some localities now set No. 12 as minimum for all wiring.

The important point is advance planning for tools and appliances that you are reasonably likely to want in the future. It is much easier to install the wiring for them when the overall wiring work is being done than later. But, if you must add such wiring to an old house there are ways of making the job easier. You'll find them covered in detail in later chapters on Surface Wiring and Modernizing.

WIRE SIZES AND CAPACITIES FOR 115 VOLTS*

WIRE SIZE	MAX. FUSE AMPS.	DISTANCE (FEET) ONE WAY FOR AMPS AND WATTS					
		5A 575W	10A 115W	15A 115W	20A 2300W	25A 2875W	35A 4025W
14	15	90	45	30			
12	20	140	70	47	35		
10	25	220	110	75	60	45	
8	35	360	175	125	90	75	55
6	45	560	280	190	150	120	85

A FEW WORDS ABOUT THE NATIONAL ELECTRICAL CODE...

THE FIRST nationally recommended wiring code, or set of rules, was published in 1895 by the National Board of Fire Underwriters. It became the basis of the National Electrical Code that was drawn up two years later through the efforts of insurance, architectural, electrical, and related groups, all of whose delegates voted for adoption or approval of the Code. Today, expanded and amended many times to keep pace with the ever-greater use of electricity, the National Electrical Code is the standard of the National Board of Fire Underwriters and recommended by the National Fire Protection Association.

The Code aims to assure safe wiring materials, devices, and practices. It is not a wiring instruction manual. It specifies the correct materials to use under various conditions, and it defines the correct methods of using them. It does not have, in itself, the force of law. But when a community adopts the Code as a part of its local regulations, it acquires that force. Often, too, the National Code is combined with a local code which may be more restrictive. Also, local codes vary in the extreme. In a sample sixty-mile range, for example, one city banned homeowners from doing any wiring work. Another not only permitted them to do it, but offered a free pamphlet to help them avoid errors. Another permitted them to do it after paying a considerable fee, and still another had no regulations whatever. So check your local code.

Perhaps because of the many national and local code combinations, the National Code is often misquoted. You may hear, for example, that the National Code prohibits the homeowner from doing his own wiring. Actually, the National Code doesn't even touch on the subject. In general, its a good idea to have a current copy of the National Code on hand if you plan any wiring work. And, of course, a copy of the local code should be with it. The National Code is available from the National Board of Fire Underwriters to those whose work requires it. If you are planning a wiring job, however, your best source of a copy is likely to be the electrical supplier from whom you buy your materials.

The code book, itself, is a little under 6½ by 3¼ inches in size, with about

The National Electrical Code doesn't tell you how to do wiring, but it specifies materials for the job and safe methods for using them.

450 pages. It is indexed by "Articles" and sections of articles, all numbered for easy location. For example, if you want to tape a splice according to the Code's recommendations, you can look in the index under "Splices and taps." Listed under that heading you will find "General Provisions. 110–14." The article is 110, the section 14. As the articles run in order, it is a simple matter to locate the specific one, then the section. If you are accustomed to looking up indexed items by page number, avoid confusing the two systems. Ordinarily, the chances are you can find the subject you want within seconds. When a subject seems to be missing look under other possible classifications. If you still cannot find it the chances are it is not covered by the code.

Don't let the thickness of the printed Code book mislead you into feeling that looking up a particular item will be complicated. A considerable portion of the book is devoted to special problems (and their solutions) likely to be encountered in industrial and commercial work and other nonresidential wiring.

In general, it is worthwhile for the prospective homeowner-electrician to read through the Code's rules for residential work. The language is plain and clear-cut, and the recommendations logical. And there's an equally important reason for familiarizing yourself with the National Code: you won't be mislead by the various bans and restrictions incorrectly attributed to it.

CHAPTER ONE

TOOLS FOR WIRING

IF YOU have an average fix-it tool kit, the chances are you already have most of the tools you need for house wiring. But a few special extras may be required, depending on the type of wiring you'll be doing. For example, you'll need a reel of "fish tape" if you plan to snake new wiring through old walls. And you'll need a conduit bender if you plan to use conduit. If you plan simple jobs like adding a few outlets, however, the chances are you can do it with everyday tools.

Pliers. Ordinary utility pliers (sometimes called mechanic's pliers) can usually handle most wiring jobs, and are probably the type you have learned to use. Electrician's pliers are, of course, the professional's choice for wiring work, but are not vital to your kit. Their broad jaws get a firm grip on wire and their side cutters do a neat job of cutting it. But, as the jaws do not close completely (in order to give the side cutters positive action) they are not suited to general work.

Small-nosed pliers are one type you should have for wiring work. They

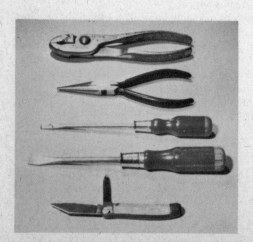

For fix-it jobs like replacing switches and outlets, all you need are these common tools (from top): mechanic's pliers, small-nose pliers for looping wire ends, electrician's screwdriver, standard ¼" screwdriver, and pocket knife. If knife does not have punch blade, add an awl.

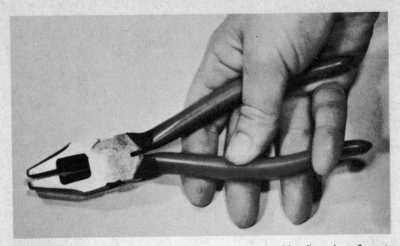

When doing wiring work, hold pliers with little finger on inside of handle and use finger to open pliers.

make it very easy to form the small loops in wire ends to fit terminal screws, and they come in handy for reaching into tight quarters in outlet or switch boxes. They also have a wide variety of other uses for odd jobs around the house or in radio, electronic, or ignition work.

Insulated handles on pliers are desirable from the standpoint of comfort.

For bigger jobs involving armored cable, conduit, or raceway, you'll need these extra tools: brace and bit, hacksaw, compass saw to make wall openings, electrician's pliers for easier splicing, Yankee push drill to start screw holes for mounting outlet boxes on wood framing.

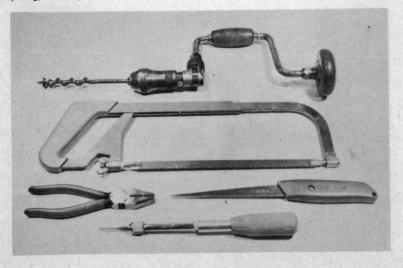

But don't ever consider them as sufficient safety factor for handling live wires. A pinhole in the plastic and a drop of perspiration could prove disastrous. Stick to the practice of working on wiring only with the current definitely off.

Screwdrivers. Favor plastic-handled screwdrivers for comfort and durability. If you must do any work in the entrance panel after the power lines are connected, the insulated plastic handles are safer. Even with the main switch *off*, there are live connections on the power-line side of the switch.

You will find an electrician's screwdriver handy both in your wiring work and on general fix-it jobs. It has a long, slender blade with the flat portion of the blade tip the same width as the round shaft. This lets you work in tight spots and—in fix-it work—drive screws that are deeply recessed. Pick one with an insulated blade shaft. One of your screwdrivers should be rugged enough to take a hammer tap on the end of the handle. This is regarded as poor tool technique but it's probably the most widely used method of opening a "knockout" in an outlet box. Most good plastic-handled screwdrivers can take it without damage.

Hacksaw. For cutting armored cable, conduit, and other types of wiring material, a hacksaw is a valuable tool to have in your kit. Use a fine-toothed blade, as most of the metal you'll be cutting will be thin. A coarse-toothed blade is likely to catch and chip teeth when used on thin stock.

Metal Snips. Quicker and handier than the hacksaw on flexible armored cable, metal snips are less likely to cut into the wires when all you want cut

Essential tools for extensive electrical work (from top): flexible rule with lock button, neon test light, aviation shears for cutting armored cable and sheet metal, pipe reamer (left of shears) for smoothing sawed conduit, special Simonds mower blade file for smoothing inside of small hacksawed raceway.

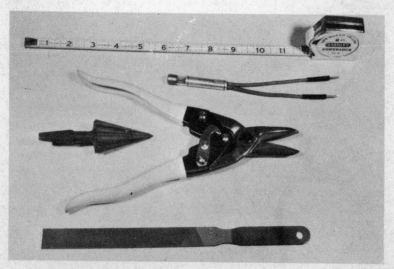

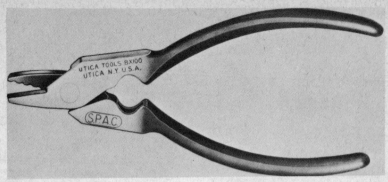
Special plier-cutter tool for cutting and working BX armored cable.

is the armor. Some electrical suppliers also stock a somewhat similar tool designed specifically for cutting armored cable. Metal snips like Stanley's aviation snips are likely to have a wider range of uses after the wiring work is finished.

Soldering Tools. If electricity is already available, and the section of the wiring you are working on is not yet connected into the system, an electric soldering iron or soldering gun is the best tool for the job. If electricity is not available, your best bet is a torch and an old-fashioned soldering iron.

In selecting any of these tools, check their heating capacity. Favor a large one over a small one, as the wire ends to be joined must be heated by the soldering tool to a temperature high enough to melt the solder. This assures adhesion of the solder directly to the wires, and a thorough bond between the joined pieces. You must be able to touch the solder to the heated wire and melt it so that it flows into the windings of the splice and adheres to all parts.

Insulation Stripping Tools. There are various multipurpose plier-type tools designed both for cutting and bending wire and for stripping the insulation from it. Your electrical supplier stocks at least one. This is a handy tool for extensive wiring jobs like an overall house installation. But for the small job or occasional electrical repair the pocket knife is hard to beat.

The Circuit Tester. This is a somewhat specialized tool that is most likely to figure in checking wiring after it is completed to determine if a current is present. In its simplest form it is a small neon glow lamp with a pair of wire leads extending from it. Insulated ends on the leads allow you to hold them with their protruding metal contact tips against wires carrying current.

If, for example, you want to know whether there is danger in touching the outer shell of a kitchen appliance while turning off a faucet (remember, your wiring is connected to the water pipe for grounding), you can place one of the tester leads against the appliance, the other against the faucet. If the test light glows there is a "potential," or current from one to the other. This is often the case with older appliances that are not connected to the outlet with grounded plugs. If you encounter this condition, turn the plug around so the prongs are in opposite positions in the outlet, and repeat the test. Usually you will find that the test light does not glow. Then mark plug and outlet with paint so the plug can be replaced in the outlet in this position.

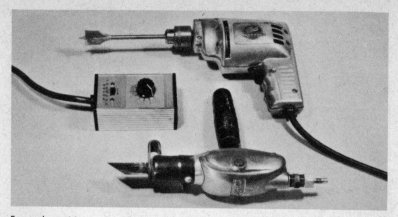

For work requiring considerable boring, save effort with a power drill. Hacksaw attachment comes in handy for cutting raceway and conduit. Use it with motor speed control like portable unit shown.

Woodworking Tools for Electrical Jobs. A hammer, a hole-boring tool, and sometimes a chisel will be useful for wiring work. You'll need the hammer for driving the staples that hold armored cable to house structural members and for a variety of jobs including the tapping open of knockouts in outlet boxes. If no electricity is available, you can use a bit and brace to bore the holes in studding and general framing through which cable will be led. Naturally, a power drill speeds the work if there's current available. The chisel comes in handy in conduit work (where you'll also need a keyhole saw) and in other wiring where a recess must be made in wood.

Insulating Tapes. The most widely used type of insulating tape is usually referred to as "plastic tape" and commonly includes a vinyl layer. It is very thin, but insulates better than older types many times its thickness. The fact that it is so thin makes it very handy to use in small outlet and switch boxes where thicker tape would result in a bulky splice. Although this type of tape is suited to almost all your wiring needs, it can't be used at extremely low temperatures, as in an unheated building in cold weather. A special form made by Johns-Manville, however, can be used at temperatures as low as 10 degrees below zero without cracking.

In extremely damp or wet areas, rubber tape is used, as it actually vulcanizes itself together at overlap areas to form a continuous sealed tube. It is thicker than the plastic tape but not as tough. To protect it after it has been applied, give it a covering of friction tape, which has ample resistance to the usual types of mechanical damage.

Solderless Connectors. These are made in several types to reduce the need for soldering splices. In their handiest form they consist of a hollow plastic cap with a spring-like wire winding inside to serve as a conical thread. To connect a pair of wires you simply insert them inside the thread and give the cap a few turns. The wires are twisted together in a gentle spiral by the motion and firmly locked together by the conical thread. They provide a safe and convenient way of doing electrical work without the delays of soldering.

CHAPTER TWO

TYPES OF WIRES AND WIRING TECHNIQUES

ELECTRICITY IS carried from its source to the point where it is used by wires covered with insulating material. Copper has long been the most common type of wire used, although aluminum is used in certain types of wiring, such as long-distance transmission lines, because of its lightness and strength.

Wire Sizes. The size of each wire is determined by the cross-sectional area measured in circular mils. One circular mil is equal to the area of a circle whose diameter is 1/1000th of an inch. Wire sizes are designated in the United States by the American Wire Gauge (AWG), also called the Brown and Sharpe Gauge (B&S). The sizes range from the smallest at 40 (B&S) and even 60 (AWS) all the way up to 0000; the smaller the wire, the bigger the number.

Sizes from 50 up to 20 are very small and, except for such uses as door-bell circuits, are seldom employed as power wires. Number 50 is about the diameter of a human hair and weighs about 1 pound per 60 miles. Wire of this type is used in delicate electrical instruments and electronic work.

Sizes 18 and 16 are used in stranded form in common lamp cord, which is discussed in detail below. Stranded wire size is the overall size of the group of strands, not the size of the individual strands. It corresponds in diameter to solid wire of the same size designation.

Numbers 8 and 6 are used in either solid or stranded form and are normally employed as service-entrance leads to homes and small buildings, and for heavy power circuits. Wires larger than 6 are for carrying heavy power loads from 55 up to as much as 320 amperes, in weatherproof 3/0 size, and are used in some instances for service-entrance cables and for industrial purposes. Number 4/0, used for the same purpose, is almost ½″ in diameter.

Cords. Wires that are used in the home to operate appliances, lighting fixtures, etc., are called cords, and are classified according to how they are used and constructed. The simplest and most common type is *lamp cord*, which consists of two stranded copper wires covered by rubber or plastic insulation. These are usually available in brown or ivory-colored insulation, but

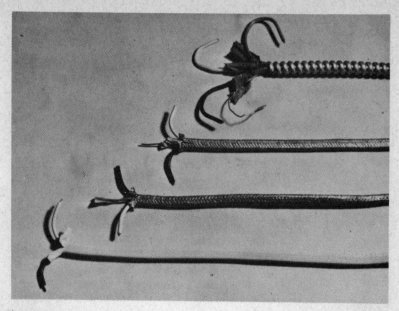

These are the basic types of cables (from top): 3-wire BX with bare grounding wire, also available in 2-wire form for general wiring; Romex nonmetallic cable with two wires plus bared ground; Romex with two wires, center filled with special twine in place of ground wire; underground fused 2-wire cable for wiring driveway lamp posts, etc.

are manufactured in several other colors as well. Another type of lamp cord contains the same double wires with rubber or plastic covering, but is wrapped with cotton or silk braid to match upholstery, rugs, or other household furnishings.

The second type of cord is *heater cord*, used with heat-producing appliances such as toasters, irons, hot plates, portable heaters, etc. This cord consists of two copper wires covered by rubber insulation, a layer of asbestos fiber and covered with heavy cotton braid.

The third type is *heavy-duty cord*, which has two or three conductors protected by fiber plus rubber or plastic insulation. In the three-conductor type of heavy-duty cord, each wire is covered by a different colored rubber insulation, usually one black, one white and the other green. This color-coding tells you that the black is the "hot" wire, white is the "neutral" wire and the green is the "ground" wire. Three-wire cord is used, for example, with portable power tools; the black and white wires are the conductors while the green wire is connected to the metal case of the tool, thus grounding it. This minimizes the possibility of electric shock if an internal defect develops in the tool. For more detailed information on cords, see Chapter Eight.

Cables. To carry the heavy load of electricity from the service entrance of a home to the distribution panel where the fuses or circuit breakers are located, large multiple-strand wires, or cables, are used. From the distribution panel the various circuits are carried by several types of conductors approved by the National Code for use in residences and farm buildings. There are three basic types of approved multiple-strand wires used in interior wiring, not including the surface wiring covered in another chapter.

Flexible armored cable. One of the most widely used and one of the easiest to install, this cable is available with either two or three insulated wires with an outer wrapping of tough paper and enclosed in a spiral-wrapped galvanized steel armor. The armor is wrapped in such a way as to be flexible. You are likely to hear this type of cable referred to as BX, which is actually a trade name. But there are other trade names for similar cable, such as Flexsteel. (By whatever name you buy this or any other type of wiring material, make certain that it carries the UL approval.) Armored cable of this type can be used in general wiring, concealed or exposed, but it may not be used in damp locations or out of doors.

Nonmetallic sheathed cable. Known by such trade names as Romex, Cresflex, etc., this is another interior cable for general dry-location wiring. It is available with either two or three wires, like BX, all enclosed in a specially treated fabric braid, and is one of the more inexpensive forms.

Plastic-covered cable. This is the type of waterproof cable to use in damp locations, out of doors, and underground, as in wiring a gate light, or leading a power line to an outbuilding. The official terms for the different cable types are given shortly, but the best bet when you buy is not only to specify the type of cable but the purpose for which you intend to use it. Your electrical supplier is likely to know if local codes require a specific variation.

Thin-Wall Conduit. A light steel tubing which is bent with a special tool to conform to the path the wiring must follow, thin-wall conduit is not made with the wires already in it, as the cables described previously. It is installed somewhat like plumbing, then the wires are pulled through it. As it is not flexible, and must be bent to shape on the job, quite a bit more work is involved in using it, but codes in some areas require it. It is made in 10′ lengths that can be joined with threaded couplings that clamp over the ends of the conduit, which need not be threaded for the purpose.

Where the conduit enters an outlet box or other type of box (always metal), a connector attaches to it in the same manner. The threadless end of the connector is first fitted over the conduit and tightened. Then the connector is inserted through the box knockout hole and fastened with its own lock nut. After the conduit is installed and connected to the boxes the wires are pulled through it, leaving about 8″ of wire inside the box at the end of each run to allow for connections.

Conduit is cut with a hacksaw and reamed at the cut ends with a pipe reamer to remove any sharp burrs left by the sawing, so as to remove any chance of insulation damage. The wire size in the conduit is matched to the capacity of the circuit.

The only other type of wiring you are likely to use in home installations is surface wiring, covered in a separate chapter under that heading. This is especially suited to use in old houses when the wiring is to be modernized, as it entails a minimum of wall openings.

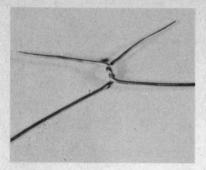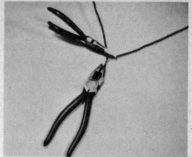

Simple pigtail splice is made by twisting ends of two wires to be joined as shown at left. (Soldering wire used in these illustrations for clarity.) When using wire larger than No. 14, use one pair of taped pliers to hold wires (protecting insulation), another pair, preferably electrician's pliers, to do the twisting (right).

How Cables are Designated. The types of cable just described are available at all electrical suppliers in areas where they are commonly used. They are designated by letter names. Armored cable such as that carrying the trade name BX, for example, is referred to in the national code as Type AC. But if you ask for BX you can be sure the electrical supplier will know what you mean, as the trade names are as widely used as the letter designations, perhaps more.

Nonmetallic sheathed cable like Romex is designated as Type NM. When it is designed for use in wet or damp locations it is designated as Type NMC, which, of course, can be used also in dry locations.

The oldest form of underground cable is designated as Type USE (Underground Service Entrance). A newer type designated as Type UF (Underground Fused) first got its code recognition in 1953. This can be used in much the same way as type USE (underground), except that it must be protected by fuses or circuit breakers at the starting point. For instance, if you want to run underground wiring from your house to a separate garage, you must provide fuses or circuit breakers for the wiring at the house end. In multiconductor types with two or three wires, Type UF looks exactly like Type NMC. Some brands also have dual Underwriters' approval for use as Type UF-NMC, which means they can be used for interior wiring or underground runs. If the various letter designations confuse you, explain to your electrical supplier what you want to do, and he will be able to specify the correct cable for the purpose. He is also likely to know of any local code restrictions on cable types.

BASIC WIRING TECHNIQUES. Most of your splicing will involve the joining of either two or three wires together. When you wire a switch, for example, you will be joining the two white wires, as only the black ones connect to terminal screws. The code requires that all splices be mechanically and electrically secure before soldering, and also that they be soldered. In other words, the joint has to be strong and it has to make good electrical contact before the solder is applied. The solder locks it. Completed, it must be as good a conductor as the wire.

The Pigtail Splice. To make this splice, bare about 2″ of the wire ends and scrape them clean. With the bared ends close together and parallel (insulated sections lying side by side) grip the bared sections as close as possible to the insulation. Then twist the bared ends together like rope. On heavier wire like No. 12, a second pair of pliers comes in handy for the twisting. Five or six turns are about enough. Turn the remaining ends back alongside the twist so they won't tend to poke through the tape later. This type of splice is not intended to take tension, but you can test your work with a light pull on the wires. Then you are ready for soldering.

The Western Union Splice. This is a stronger form that may take slightly longer to make but it can take tension. Bare 3″ of the wire ends and scrape them clean. Give each wire a right-angle bend about half an inch from the insulation and hook the two bends around each other. Hold the overlap with pliers and bend the free bared end of one wire to start wrapping it around

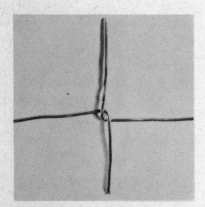

Three steps in making the Western Union splice: First, join wire ends by making an L-bend in each wire (left). Twist one end (in this case the right-hand one) around the main part of the other wire (bottom left). Complete splice by wrapping other end around second wire (bottom right).

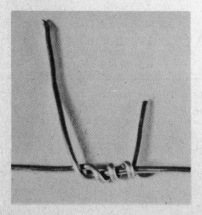

Bunch splice, used when more than two wires are to be combined, starts like this (left). All wires are twisted together like strands in a rope. Completed bunch splice is shown at right.

the other, working toward the insulation. When the first one is wrapped as far as it reaches, do the same with the other, working in the opposite direction. The finished job will take plenty of pull.

The Bunch Splice. The name often given to the pigtail splice when it joins more than two wires, this splice is made in much the same way as the usual pigtail. Twist the three wires simultaneously so they will fit together like the strands in a rope. If two strands are twisted first, then over-wrapped with the third one, the splices will not be as secure.

The Tap Splice. This is the type to use when a wire is to be joined to the run of another wire. Strip about a 1″ section of insulation from the running wire and scrape it clean. This will be a bared section with insulation beginning again at each end. Bare about 3″ of the joining wire, and scrape it clean. Now, starting at one end of the 1″ bared section of the running wire, wrap the joining wire toward the other end. In the usual wire sizes the 3″ bared length will wrap just about all the way along the bared inch. That's all there is to it except for soldering and taping.

Soldering Splices. The trick in sound splice soldering is to heat the wires until the solder melts on contact with them and flows into all crevices like paint. Any good soldering iron (large is better than small) will do the trick. Although rosin-core, wire-type solder can be used for the job without additional flux, it is a good idea to use a non-acid (rosin) flux also.

Smear the flux over the splice before heating. It will melt and run into the crevices the solder will fill later. Hold the splice against the soldering iron with pliers, touching the end of the wire solder (it looks like silver wire) against the wires until you see it soften and begin to spread into the twist. Then push the solder inward against the wire to supply enough more solder to complete the bond. It is actually drawn into the crevices by capillary attraction, like water soaking into a sponge.

It helps to turn the splice over quickly during the work to get a smooth filling on both sides. Allow a few seconds for cooling with the iron removed from the wire. You can tell when the solder has hardened by a change in the surface sheen.

Taping. There are three types of electrical tape in common use. The most widely used is vinyl plastic tape, as it has the greatest insulating power for any given thickness. This allows a very safe and thorough taping job inside of small outlet and junction boxes without danger of inconvenient bulk at splices.

Rubber tape is used where moisture is a problem. When this tape is spirally wrapped around a wire, with tape edges overlapping, it actually vulcanizes itself together, forming a completely waterproof hose. (If you need a small piece of rubber hose in an emergency you can make it this way. Wrap a round pencil with cellophane or similar plastic, then wrap on the rubber tape. Slide it off and you have your hose.)

Friction tape, the old standby, is now used only on outside wiring, as far as insulating taping is concerned. Some of the top quality types can be used alone for insulating purposes, but generally are not because of their porosity. Friction tape is widely used over rubber tape, however, to protect it from abrasion or mechanical damage, as the rubber used is very soft.

If you are doing tape work at low temperatures, as in an unheated garage in winter, you may require a special low-temperature type of vinyl tape, as the standard form cracks when flexed in very cold weather. The special form can be used at temperatures as low as 10 degrees below zero.

Taping technique. Start your taping a little more than a tape width back on the insulated portion of the wire and wrap on to the splice, pressing the tape down firmly into hollows. Wrap across and beyond the start of insulation at the opposite end for a distance equal to that at the start. Then tape back again, spiralling the wrapping in the opposite direction, so the spirals criss-cross.

First stage in making tap splice (left) shows tapping wire with turn around running wire. Splice is completed (right) by simply continuing twist of tapping wire around running wire. All splices are soldered and taped before using wires.

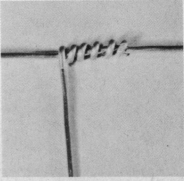

SOLDER

INSULATING
TAPE

Sears, Roebuck & Co.

Tape all splices after soldering. If soldered sections are rough, give them an extra layer, then wrap from well back on insulation onto and across bared section to insulation, then back again.

In all wrapping, allow an ample overlap of the tape edges. In the past, three to four layers of rubber tape were used followed by two or three layers of friction tape. Now you can do it all with about three or four layers of plastic tape. In general, if the plastic wrapping is a little thicker than the insulation on the wires, you're on the safe side. Be sure the finished job is smooth, with no loose turns or ends that can catch and cause unwrapping. (You can get the technical insulating rating of both wire insulation and tape for precise matching, but it's rarely done.)

Solderless Connectors. The splicing job can be simplified with wire nuts, which can be used anywhere that the wires are not under strain or tension. In insulated form they can be applied in seconds.

Bare the wire end just far enough to fit all the way up into the cap of the connector, and be sure the wires are scraped or sanded clean. Then hold the bared wire ends together, pointing in the same direction and screw the solderless connector on them. A spring-like metal thread inside the connector tapers toward the upper end so as to press the wires tightly together as it tightens.

In taping bunch splice, start tape on one branch and run onto bare section (left). Tape a little way beyond the end of bare wire (right), then back onto other branches. Fold over excess at end of splice and use a separate piece of tape to make an extra turn or two. Rubber tape is best for complex splices like this one.

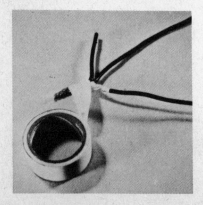

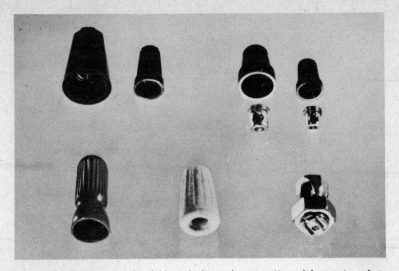

Types of solderless connectors available to the home electrician: Upper left, two sizes of connector that simply screws on wire ends. Upper right, two sizes with connector containing set-screw which is tightened on wires, then cap is screwed on. Bottom left, flexible plastic screw-on connector; center, screw-on ceramic type for high temperatures; right, brass, heavy-wire connector, tightened with wrench. This one must be tape wrapped.

If the ends of the wires have been bared to the proper length, no taping is necessary. If bare wire extends below the connector (as when too much wire has been bared), tape the exposed portion.

An uninsulated and somewhat lower priced form of solderless connector is also available. This type saves the trouble of soldering but requires tape. It is still a time and work saver.

A Splicing Tip. Before starting on splicing work for the first time, you can learn the best wire-handling techniques by practicing a little with a few lengths of wire solder. This is soft enough to splice without the aid of pliers, providing an easy way to see just how the wires fit together during twisting in each type of splice. And when you are through with your practice you can still use the scrap pieces of solder for soldering. As the wire solder compares in size to rather heavy copper wire (No. 12 or larger) it will also give you a good idea of the space the splices will take up and the length of wire needed to make a given number of turns around another piece.

Sears, Roebuck & Co.

To use screw-on connector, bare just enough of wire ends to extend all the way into the cap. Place wires side by side and screw connector onto wires.

CHAPTER THREE

HOW TO INSTALL HOUSE WIRING

BEFORE STARTING any wiring work, especially if it is of a major nature, find out if there is a local wiring code. You can usually do this by phoning the building inspector or your power company. If there is a code obtain a copy before undertaking any electrical job on your own. Often, communities use the National Electrical Code as the general basis for their regulation, but add certain rules especially applicable to the area.

The local attitude toward non-professionals working on wiring, the homeowner, for example, varies widely. Some large cities prohibit the non-professional from doing any wiring work at all. Others not only permit it but supply pamphlets and other data to help the non-professional avoid errors. Some require that a permit be obtained and that inspections be made by an official of the building department at various intervals as the work progresses. These are among the important points to know before you undertake your wiring job. The rest is mainly a matter of careful work.

On any complete job of house wiring contact your power company in advance. In most cases they will send a representative to check over the location. He will be able to tell you what materials you must supply and what materials and work the power company will furnish in connection with your service entrance. He can also advise you on the best location for the entrance so that entrance (overhead) wires will not traverse someone else's property or otherwise create problems.

THE SERVICE INSTALLATION. Three separate wires enter your house in a modern service installation, one with black insulation, one with red insulation, and one bare. The bare one is called the neutral, or ground wire. The other two are called "hot" wires. Actually all three are hot in the sense that they are carrying current, and all three must be treated with equal respect as far as danger is concerned.

The bare neutral wire is connected to a metal "neutral strip" inside the service panel (you may call it a fuse box). Another wire from this strip runs to a connection on the water pipe of the house, or, if there is no water pipe, to a metal rod made for grounding purposes and driven into the earth. If the length of the water pipe is less than 10′ underground, the ground wire should be connected both to the pipe and a ground rod.

The black wire commonly is connected to the left terminal of the "main disconnect" and the red wire to the right-hand terminal. Also in the panel

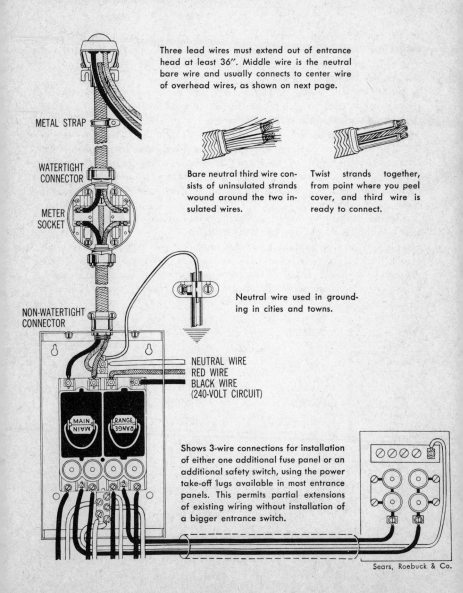

Three lead wires must extend out of entrance head at least 36″. Middle wire is the neutral bare wire and usually connects to center wire of overhead wires, as shown on next page.

METAL STRAP

WATERTIGHT CONNECTOR

METER SOCKET

Bare neutral third wire consists of uninsulated strands wound around the two insulated wires.

Twist strands together, from point where you peel cover, and third wire is ready to connect.

NON-WATERTIGHT CONNECTOR

Neutral wire used in grounding in cities and towns.

NEUTRAL WIRE
RED WIRE
BLACK WIRE
(240-VOLT CIRCUIT)

MAIN
MAIN

RANGE
RANGE

Shows 3-wire connections for installation of either one additional fuse panel or an additional safety switch, using the power take-off lugs available in most entrance panels. This permits partial extensions of existing wiring without installation of a bigger entrance switch.

Sears, Roebuck & Co.

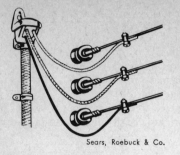

To connect wires from entrance head to overhead wires, the Code requires that the entrance head should be installed above the top insulator of incoming power wires, and that drip loops should be formed in each conductor to prevent water from entering system.

Sears, Roebuck & Co.

are the starting terminals of the separate pairs of wires that will be the individual circuits of the house wiring, each with its own fuse. There are also terminals for 3-wire circuits such as those required by electric ranges and other high-wattage appliances.

About the Two Voltages. The operation of the 3-wire system is simple. If you were to connect a voltage-measuring device to the black wire and to the neutral wire (the neutral wire is covered with white insulation in all interior wiring), you would get a reading of 120 volts. If you were to connect the same instrument to the red wire and the neutral wire, you would get the same reading. But if you connect it to the black wire and the red wire (called the "hot" wires), you would get a reading of 240 volts. These, then, are the basic connecting arrangements used to supply either voltage, according to the needs of the appliance.

The diagram supplied with the service panel should make it clear exactly which terminals should be used to start circuits of either voltage. If it is not clear to you ask your supplier to explain it. And, if you have any doubt whatever as to whether your connections are correct, pay a professional electrician to check them over before you have your power connected. In localities where inspections are made, the inspection will include this point. The con-

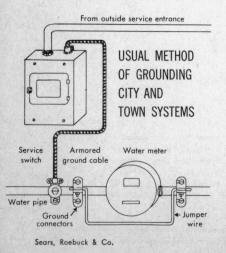

From outside service entrance

USUAL METHOD
OF GROUNDING
CITY AND
TOWN SYSTEMS

Service switch · Armored ground cable · Water meter · Water pipe · Ground connectors · Jumper wire

Sears, Roebuck & Co.

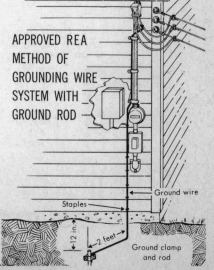

APPROVED REA
METHOD OF
GROUNDING WIRE
SYSTEM WITH
GROUND ROD

Ground wire · Staples · 12 in. · 2 feet · Ground clamp and rod

First step in cutting BX armored cable is to make a sharp bend, squeezing cable enough to buckle armor.

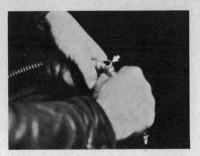

Next, twist cable against direction of spiral to spring out buckled section of armor so cutting tool can be slipped under it.

Slip cutting tool under raised section and cut through armor, taking care not to damage enclosed wires.

With large jaw recesses of cutting tool, reshape end of armor that was buckled in severing the cable.

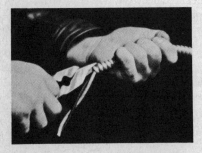

Final turn of spiral is shaped to conform with those behind it with jaw tips of cutting tool. Cable is now ready for bushing.

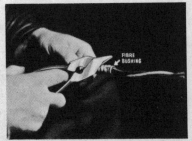

Kelsey-Hayes Co.

Fiber bushing is slipped over wires and pushed back to edge of armor. If wires are covered with paper wrapping, remove first.

nections are not at all complicated, and no maze of wiring is involved, as in a
TV or radio chassis. But take no chances if you are not sure of your work.

Installing Wires in the Service Panel. The tool work begins with your first
connection. Sometimes the circuits are wired from the outlets back to the
service panel, then connected in. Other times they are started from the panel
and led out to the outlets. So long as the power is *off*, either method is practical. For simplicity we will start at the panel (before the power is connected)
and lead a typical circuit out to the fixtures and outlets.

With the cover off of the panel hold the wiring cable against it to determine
the length of wire that must be freed to run from the wall of the box to the
connecting terminals. Assume that BX cable, one of the most commonly used
types, is being installed. If you require 8″ of free wire to reach the terminals,
you will have to remove that much metal armor from the end of the cable.

Cutting the cable. Bend the cable sharply at the 8″ mark and squeeze the
bend inward until the armor buckles. You can feel it buckle and you will note
that one of the spiral turns at the center of the bend bulges outward. Now
grip the cable with your hands on both sides of this buckling point and twist
it against the direction of the spiral winding until the buckled turn of armor
moves out enough to slip the jaw of a metal snip between it and the wires,
which are exposed at this point. Snip through the strip of armor and you
can slip the severed section off the end, freeing the wires. You can use a
hacksaw to sever the armor instead of metal snips, but it takes longer and
calls for care to avoid cutting into the wires.

If a jagged point remains on the severed end of the armor, trim it off so it
will not cut into the wires' insulation. If the end of the armor is distorted (not
likely), it is easy to reshape it with pliers. After a little experience you will
be able to snip off armor without need for trimming or reshaping. If you are a
regular user of fix-it tools, you can probably do it the first time.

Attaching bushings. With the armor slipped off the wire end, remove the
water-repellant paper wrapping around the wires back to the point where
the armor begins. Next slip a fiber bushing over the wires and push it back
into the armor. These are standard items sold by the box at electrical suppliers. A small collar around one end prevents them from slipping all the way
into the armor. Never make a BX connection without using a bushing. The

Two important steps in attaching armored cable to outlet box: (left) insertion of fiber bushing
to protect wire; and (right) attachment of connector to hold bushing in place and connect
cable to outlet box.

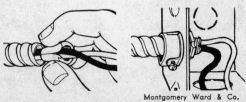

Montgomery Ward & Co.

Fiber bushings for BX cable ends can be purchased by the bag. Batch here would be enough for the average home wiring job.

bushing prevents the wire insulation from being chafed or cut by the end of the armor.

Installing the connector. With the bushing in place slide a connector (made for BX) over the wires and on to the end of the armor. The threaded end should face outward, toward the wires. Seat the connector snug against the armor end and the bushing inside it, then tighten the setscrew that holds the connector to the armor.

Opening panel "knockout." Now you are ready to open a "knockout" in the metal case of the service panel. (The same type of knockout is used in all metal wiring boxes, for outlets, switches, etc.) To open it, note the location of the small points at which it is attached to the metal around it. Then place a screwdriver against it midway between these points and give it a sharp hammer tap. This partially opens the disk-shaped knockout. Grip the opened disk rim with pliers, give it a little twist, and it comes free, leaving a hole that exactly fits your cable connector. *Do not* open any knockouts that are not to be used. On service panels and all other metal boxes used in your wiring you will find knockouts in a variety of locations to permit wiring to enter and leave along the most direct path.

To connect wires to service panel, insert the connector through the hole left by the knockout and screw the locknut on its threaded inner end. As these nuts must be used in many boxes much too small for the use of a wrench, they are made with notches around their perimeter. To tighten them, insert a screwdriver blade in a notch and either pry the nut around or hammer-tap it around to tighten it. Tightening takes only a few seconds but it is important. The armor of the cable, like the white wire inside it, is grounded. Its connections to boxes should be both mechanically and electrically sound. Then, if the insulation on a black wire should be damaged, or a black wire splice poorly insulated so that the black wire contacts the inside of the cable armor or box, a fuse will blow and warn of the trouble. Poor connections between armor and box could result in sparking or arcing in such a case. This can sometimes take place without blowing a fuse, yet create sufficient heat to cause a fire. While such occurrences are not frequent it pays to take all precautions to prevent them.

Extra-long bit for power drill aids in boring holes in joists to carry armored cable through house.

Right-angle attachment enables you to work between joists with power drill. Flat-surfaced bit assures drilling stability.

After boring holes in joists, thread armored cable through, taking care to remove all kinks or sharp bends.

These cables lead into wall above, holes having been bored *downward* through sole plate of upper wall. From basement point of entry, they will go to power source.

HOW TO RUN WIRING THROUGH THE HOUSE. Starting from a basement service panel, the BX usually emerges from the top of the panel and runs to the ceiling. If the joists (beams) are exposed, there are three ways in which the cable may be run:

(1) If the cable must run crosswise of the joists, it may be led through holes bored in them. If you have a power drill (and a current supply for it), this is the easiest system. If you must work with hand tools, you can still do the same job, but with some bit-and-brace effort.

(2) You may also nail a board under the joists and staple the cable to that.

(3) If the cable runs parallel to the joists, it may be stapled to their sides.

The cable must be supported at spacing not greater than every 4½". (For overhead runs closer spacing is better.) It must also be supported within a foot of every box connection. Where the cable is bent around a turn it should be of such radius that if it were a complete circle it would be at least ten times the diameter of the cable. This eliminates any chance of buckling the armor. Your first bend is likely to be where the cable changes direction from vertical to horizontal (above the service panel) for its run across the ceiling joists.

Leading Wiring from Floor to Floor. When wiring runs from one floor to another, as from the basement to the first floor, it must, in most cases, emerge inside an upper wall in order to complete its run to outlets or fixtures. In new work (a house under construction) this is done simply by boring a hole downward through the sole plate (wooden base of the wall framework) to emerge through the ceiling of the floor below. In doing this, however, you must avoid boring into a joist, as you would have to bore all the way through it. Aim to have your hole so located as to come out between the joists below. In old work, as in improving old house wiring, the hole is also bored from above, but by a different technique described under Modernizing Old Wiring, in another chapter.

Once your cable has been led to the upper floor, it is run through holes in the studs (wall posts) just as it was run through the joists. With most hand or power boring tools you will have to bore these holes (in both cases) at a slight angle, as the boring tool will not fit between the framing members to bore straight. This gives a wavy path to the cable, but does no harm.

Nonmetallic Cable and Conduit. The same general procedure described for BX applies to nonmetallic cable like Romex, except that the armor-cutting and removing is not required as the cable is not armored. Different connectors are used, of course, made specifically for the cable being used. It must be mounted with special straps, however, never with staples.

Conduit must be bent with a conduit bender to take the curves that lead it from vertical to horizontal runs, and it must be cut, fitted, and connected with couplings. It is held to the framework by special straps. After the conduit is mounted and connected to the boxes, the wiring is pulled through it. In most areas, however, you are permitted to use cables for wiring, saving greatly on time and work.

THE CARPENTRY AND MECHANICS OF NEW WIRING. The methods of cable and wire connecting that follow apply to the work you will do in all types of standard boxes, whether for outlets, switches, or fixtures. And you need not be wiring a new house in order to use these methods. They will be used in finishing an expansion attic or a basement game room (which we may assume will not be in a damp basement). And they will be used in modernizing the wiring of an old house.

Outlet Boxes. There are several types of outlet boxes, though the ones you will use most narrow down to two types:

The rectangular outlet box is the one in which you will mount your switches and outlet receptacles as well as various types of wall fixtures. These can be "ganged," or joined, to form a larger box to contain more outlets or perhaps a pair of outlets and a switch. To join two boxes, remove the screw-held side wall from each, fit the open sides together and replace the screws.

The junction box is the other type you are likely to use frequently. In its

Standard types of boxes used in house wiring are shown at left. From left: standard switch box, surface-mounted box, junction box. All have assorted cover plates to match, depending on use to which they are put. Some types of knockouts (right) can be removed by merely inserting a screwdriver in slot and prying knockout disk free.

type of cable to the desired location and connect it to the box, running the wires to the proper terminal screws of the bulb socket cover. If this is a pull-chain type no other switch is needed. A variety of small holes are punched in the box to permit fastening directly to the ceiling joists or other framing, or to assorted metal brackets made for the purpose when direct fastening is not possible.

The rectangular box eliminates the need for the armored cable connectors described earlier, as they have built-in clamps to hold the cable. This saves both work and money. You will still need the connectors, however, for other types. And regardless of the type of clamp or connector you will always need the protective fiber bushing inside the cable end.

Special Boxes. There are a great many special kinds of boxes made in standard form to solve special problems and meet particular wiring situations. Space limitations prevent describing them all. But, if you have a particular wiring problem involving the size or shape of a box, you can usually find the answer at your electrical supplier.

There are shallow boxes designed to fit in thin walls, narrow ones for cramped space, large ones for multiple outlets, and ceiling types designed to support chandelier loads. Simply pick the one that suits the special condition you have to meet. And, in run-of-the-mill work, use as large a standard box as you can rather than a small one. This makes the wiring easier and reduces the chance of trouble from "jammed-in" splices.

Mounting Arrangements. Boxes of all types may be mounted between studs or joists by means of brackets that extend across the space between the framing members. Many of these are adjustable so that the box need not be centered between the frame members, but located anywhere between. Mounting directly on a frame member is also a simple matter with brackets made for that purpose. All are standardized for use with the types of boxes for which they are intended. When a box must be mounted in a finished wall, as in up-dating the wiring in an old house, there are simple clamp-in devices (described in the chapter on old-house wiring) that anchor the box solidly in the plaster or wallboard. These do not require attachment to the actual wood framing. All types, as shown by the illustrations, can be installed with simple everyday hand tools. In short, it is safe to say that there is practically no house structure situation for which a mounting device is not available for electrical outlets.

When cable and box systems cannot solve the problem you can usually use raceway multi-outlet combinations. In narrow form, this type of wiring permits outlets along runs where only the width of a calling card is available, often the case under a floor-to-ceiling picture window.

Fastenings. Brackets may be mounted with nails unless the box must be attached to them first, involving danger of damage to the box in nailing. In that event the mounting should be with screws. Also, in any situation where a pull-out load will be imposed on the fastening, a screw is the logical choice.

Wiring the Boxes. When your cable reaches an outlet, switch, or fixture box, it is connected to it just as described for the connection to the service

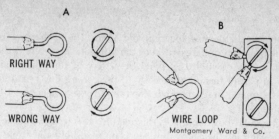

A

RIGHT WAY

WRONG WAY

B

WIRE LOOP

Montgomery Ward & Co.

When connecting wire to terminal screw, always bend loop in wire so that in tightening screw you tighten loop (A). If wire connects to terminal screw and runs on (B), bare enough of wire to make a loop as shown.

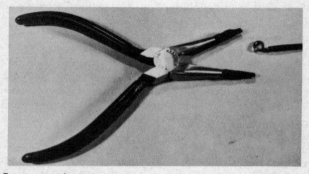

To prevent nicking wire on plier's jaws when forming loop, wrap jaws with friction tape.

panel. The same connectors (unless built-in), fiber bushings, and locknuts are used. In cutting the armor it is important also to avoid cutting the thin, flat, bare metal strip that runs parallel to the wires inside. This is built into the cable for positive grounding from one end of the run to the other. It should be bent back over the outside of the armor so that the connector will clamp it tight when it is clamped to the armor with its set screw.

The wire connections. At outlet boxes and similar ones you should allow about 8″ of insulated wire (by removing that much armor) to facilitate connecting to the terminals on the receptacle or switch. This may seem to be more than necessary, but it permits the connections to be made and the receptacle or switch to be mounted in the box without cramped wiring that might loosen connections. It also permits easy replacement of the receptacle or switch if it should fail at a later date.

Strip the insulation from the ends of the wires, leaving enough bare wire to connect to the terminal screws. This is easily done with a pocket knife, using it as in sharpening a pencil. This leaves a conical taper or the end of the insulation. *Do not* cut straight inward through the insulation. There are good reasons for this. The tapered end of the original insulation provides a better grip for tape if the wire is to be spliced rather than connected to a terminal screw. And the pencil-pointing type of knife action greatly reduces the chance of making a deep scoring cut into the wire. This scoring, which might result from cutting straight in through the insulation, makes a weak point in the wire and invites breakage when the wire is flexed. And flexing is necessary when arranging wires in the box or installing receptacles or switches.

Use the small-nosed pliers to form the loops in the wire ends which go around the terminal screws, and *always* form the loop in the direction the screw turns to tighten. If the loop is formed in the opposite direction it will tend to open as the screw is tightened, making an unreliable connection. If more wire is bared than necessary to go around the screws, trim some off with the wire cutters (most small-nosed pliers have side cutters) so as to reduce the chance of shorts.

Always connect the white wire to the chrome screw, the black one to the brass screw. Never connect them the opposite way. This applies to both outlets and fixtures. If a splice is to be made inside the box, *always connect the white wire to a white wire, black to black.* Remember, too, that all splices *must* be enclosed in an approved box. All runs with any type of cable *must* be continuous from box to box. When connecting to a switch, you will note that the switch unit has only brass terminal screws. This reminds you that only the black wire is ever connected to a switch, never the white wire. When a switch is connected into a circuit, the incoming black wire and the outgoing black wire are both connected to it, one to each of the brass screws. The two white wires remaining are either spliced, soldered, and taped, or connected to each other with solderless connectors.

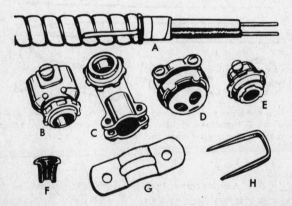

Armored-cable components: A. 2-wire cable with bare grounding wire inside armor. B. Duplex connector for admitting two cables at one knockout. C. Connector for right-angle attachment of cable. D. End fitting for changing from cable to 2- or 3-wire open wiring. E. Cable connector to hold cable in box which has no clamp. F. Fiber bushing to protect wires at cable end. G. Mounting strap. H. Staple for mounting armored cable.

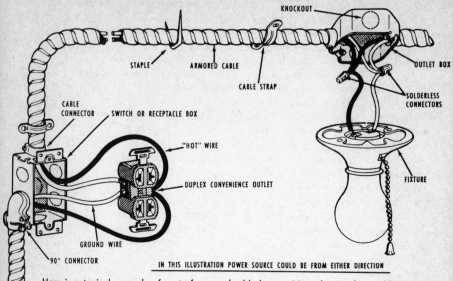

KNOCKOUT

STAPLE ARMORED CABLE

CABLE STRAP

OUTLET BOX

SOLDERLESS CONNECTORS

CABLE CONNECTOR SWITCH OR RECEPTACLE BOX

"HOT" WIRE

DUPLEX CONVENIENCE OUTLET

FIXTURE

GROUND WIRE

90° CONNECTOR

IN THIS ILLUSTRATION POWER SOURCE COULD BE FROM EITHER DIRECTION

Here is a typical example of part of armored-cable house wiring, showing how cable enters and leaves outlet and fixture boxes, and method of connecting wires to outlet and fixture. Cable can be supported on structure with staples or straps.

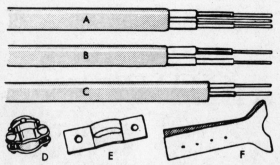

A

B

C

D   E   F

Nonmetallic-cable components: A. Cable with two wires plus ground wire. B. 2-wire cable with no ground wire. C. Plastic-covered cable. D. Cable connector. E. Mounting strap. F. Cable ripper to slit braid so wires can be separated.

The easiest way to make connections to terminal screws and to splice and tape wires (or use solderless connectors), is with the wire ends pulled well out of the box. This way, you are working in the open, not in the cramped space of the box. That is one of the reasons for the liberal wire length allowance. When you push the work back into the box, whether it is simply splices or terminal connections, form the wires into a shape that will bend easily into the box without putting a strain on the connections or splices. It is wise to arrange the wires so that when they are bent they tend to tighten rather than loosen the terminal screws. This seems like an insignificant detail, but it can do much to avoid loose connections.

Devices That Make Wiring Easier. Included here are devices that make connections simpler and others that make wiring generally more versatile. You will find these items at electrical suppliers and at large hardware stores, but not necessarily at small dealers or at the dime store.

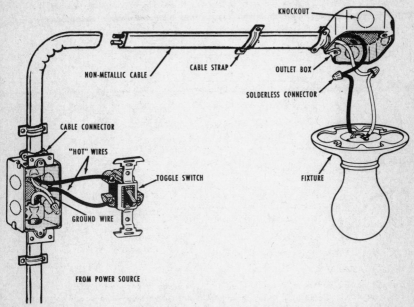

Nonmetallic cable is here wired from switch to fixture. Straps, not staples, are used in mounting this type of cable, and special connectors are used to attach cable to boxes.

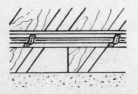

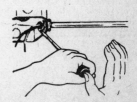

n baring wires of nonmetallic able, you can use pocketknife to lit braid, but be careful not to lamage insulation on wires.

If cable runs at right angles to joists, mount it on a 1-by-3 running board nailed across joists.

If box does not have its own clamp, and connector is used, tighten nut inside box by pushing its slotted rim with screwdriver.

Montgomery Ward & Co.

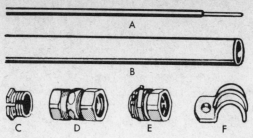

Components for wiring with conduit: A. Type TW color-coded wire used in conduit. B. Thin-walled conduit. C. Adapter for attaching thin-walled to rigid conduit. D. Coupling. E. Connector. F. Pipe strap to be used every 6' on exposed runs, every 10' on concealed runs.

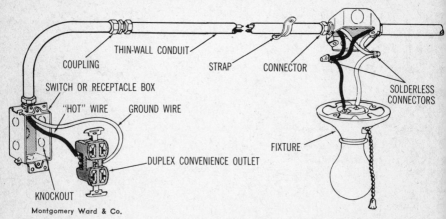

Conduit must be bent to fit installation before it is mounted. Wires are then drawn through it, special couplings are used to join lengths. All wire splices must be in boxes, not inside run of conduit.

Back-wired receptacles. You can eliminate the need for forming loops in wire ends for terminal screw connections, for example, by using "back-wired" receptacles. In the back-wired devices, the bared wire ends are simply inserted in holes in the back of the device. Then the screws at the sides (in the usual "side-wiring" position) are turned up tight. This clamps internal metal contact plates against the inserted wires and provides thorough contact. No loops in the wire are used. As to the stripping of insulation from the wire ends, stripping guides on the device assure the proper length. These guides are hollows made in the surface of the device. If you rest the wire in the hollow with the end against the end of the hollow, the edge of the device indicates the point where stripping should start. Stripped in this way, the wire has no exposed metal section when in place.

Push-in connection. Simpler still is the "push-in" type of connection. This consists merely of holes in the back of the device matched to the wire size. Push the bared wire ends (stripped according to the strip gauge) into the holes, and spring contacts make the connections and lock the wires in place. To release them for changes or alterations, a small pocketknife is pushed into a slot usually labeled "knife-point release" close to the insertion hole. The tip of the blade spreads the spring contacts that grip the wire, releasing their grip.

Interchangeable devices add versatility to your wiring by enabling you to combine a variety of devices in a single standard box. These are available in standard side-wired form (where the wire is looped under a terminal screw) or in the back-wired or push-in form outlined above. Outlets, switches, pilot lights, or other devices such as night lights may be fitted into the clamping plate, which usually takes three devices. Once in place, the devices are held there by small clamps which can be released any time the arrangement is changed. Each device is wired individually. A cover plate made to match the spacing of the clamping plate is used to finish the job in the conventional manner.

INDIVIDUAL ROOM WIRING AND LIGHTING TIPS. With the wide range of compact, interchangeable wiring devices on the market today, it is possible to have two or three switches, combinations of switches and outlets, with or without pilot lights, in the space usually required for a single old-

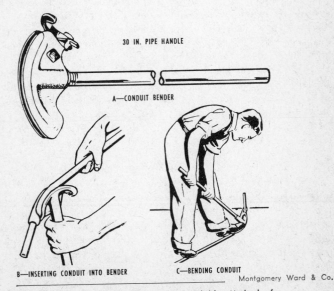

30 IN. PIPE HANDLE

A—CONDUIT BENDER

B—INSERTING CONDUIT INTO BENDER   C—BENDING CONDUIT

Montgomery Ward & Co.

Special device for bending conduit is available. Method of bending is shown in B and C.

Back-wired receptacle (left) does not require that loops be formed in ends of wires for connecting them to terminal screws as in side-wired receptacle at right. Wires are merely slipped into holes, held by tightening terminal screws.

fashioned switch. Rocker switches, almost as quiet as mercury switches, but with the added advantage that they can be mounted in any position, as they are mechanical, can have luminescent rocker buttons that give off a slight glow all night long—when exposed to any light source. For dark areas, such as hallways or staircases, switches are available with contoured handles illuminated by a small neon lamp, a safety factor in many parts of your home.

Location of Wiring Devices. Generally speaking, switches should be placed at the latch side of doors, or the traffic side of open archways. Wall switches are normally placed about 48″ above the floor. Convenience outlets, mounted 12″ above the floor, for ease in using vacuum cleaners, etc., should be spaced so that no point in any usable wall space is more than 6′ from a convenience outlet. Of course furniture arrangements, window spacings, and general architectural features will in most instances dictate a deviation from any hard and fast rule, but with these figures in mind, you can plan your wiring for the most comfort and convenience.

The Living Room. As most entertaining and family fun takes place here, careful attention to the overall lighting effect is a must. Control of illumination is easy with use of proper switches, and strategic placement of the convenience outlets permits use of radio, hi-fi, or movie projector to suit your mood. Switches placed at all entrances to the room can turn on one or all the lamps. Indirect lighting for decorative effect, or to light a display of your pet collection, can be turned on independently, or you can have any combination of lighting.

Convenience outlets should be mounted near any chair where light is required for reading, and don't neglect the desk area . . . experiment here with lamps to achieve the illumination best suited to your desk-work. Too much light creates glare, too little can make letter-writing a chore rather than a pleasure. Outlets for these areas can be planned in advance if you have a mental picture of your furniture arrangement before the outlets are installed. In an old house, where rewiring is required anyway, try out your furniture design before putting in new outlets and switches. Wall switch control of some lamps may be desirable and possible when two-circuit outlets are used.

If you want a modern wall clock in the living room, by all means install a clock-hanger outlet for it. With a recessed space for the plug, and a built-in hanger, the cord is out of sight, no dangling unsightly wire to spoil the effect.

If air conditioning is to be installed in the room, there should be a three-wire grounding type outlet near the window where the unit will be placed.

The Dining Room. In addition to the usual ceiling fixture or chandelier in the dining room, combination outlets and switches make for convenience

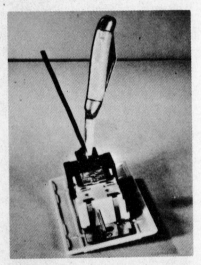

Push-in wire connections on some switches and receptacles (left) eliminate terminal screws and wire loops. Just push wire into hole; to release, insert narrow blade in slot.

Front and back view of interchangeable device that permits combining outlet, switch, and night light in one box (bottom). Pencil points to strip gauge, which shows amount of insulation to remove from wire.

in many ways. Lights that can be controlled from the kitchen, living area, and hallway are not unusual. But, with the wide range of electric portable appliances in use today, from casseroles to the old familiar waffle-iron, 20-ampere outlets should be placed within easy reach of the buffet or serving table, and don't forget outlets for the vacuum or floor polisher. For added convenience and safety, pilot or warning lights should be installed with outlets where small cooking appliances are to be used.

The Kitchen. In the kitchen electricity is truly your servant. To make the most of this, be sure that there are sufficient outlets and switches to service all the gadgets, not to mention the major appliances we all take for granted. Without proper outlets and switches you may very well let your electric tools down. Provide enough outlets, double or triple, to ensure efficient use of coffee-maker, mixer, toaster, etc. If you use many of these appliances at a time, it is wise to have more than one circuit, as some take high wattage. Here again, pilot lights will serve you well, as they are a reminder that the current is on. (The Code requires at least two small-appliance kitchen circuits.)

If a portable dishwasher is used, it will require a three-wire grounding outlet. An electric range requires a separate circuit also, with a special range type outlet.

The Bedroom. A switch at the door to light a ceiling fixture or indirect valance lighting, or both, makes for convenience and control of the lighting effect you want here. A switch with an extra convenience outlet is handy for use of vacuum or floor polisher, or for a hair-dryer. Near the bed, outlets for an electric blanket, heating pad, clock radio and bedside lamps should not be over-looked.

Wall switch controlled light at the dressing table makes the lady of the house happy, and when combined with an outlet, makes electric shaving a breeze for the master.

A well-lighted closet whether for clothing or general storage lightens many a chore. The fixture should be so placed that it does not come in contact with the contents of the closet, as this would be a fire hazard. Some fixtures have a pull chain, others have automatic door switches, such as those on refrigerators. Pass and Seymour have a new one that takes so little pressure (less than 14 ounces pressure) that it is ideal for sliding and accordion doors.

The Nursery. A switch-controlled night light can be custom designed by the home-owner. It can be combined with a wall switch at the doorway, or combined with outlets near the baseboard, if you prefer to have the light confined to the floor. Outlets for a sterilizer, vaporizor, and general lighting can be installed at the same time. Outlets that can be reached by toddlers should be the type called tamper proof or child proof, so that the little ones are safe from shock when investigating, as they always do.

The Family Room. Since this is usually the most popular room in the house, especially where the younger set like to entertain their friends, the more outlets, especially those combined with switches, the better.

When a lot of appliances, such as corn-poppers, coffee-makers and electric grills or skillets are to be used at once, not to mention hi-fi-, TV, and movie projectors, multi-outlet raceway, mounted at counter height, can take care of things, but be sure to use the baseboard type, as this carries more circuits than the narrower type usually used for this type of installation. Of course, the raceway can also be used for clean-up jobs by vacuum and floor polisher. Check with your electrical supplier when you plan to install equipment such as air-conditioners—some operate on 220 rather than 110—so be sure.

The Bathroom. A flood of light on the bathroom mirror is most desirable, for makeup and shaving. Controlled by a wall switch at the door, a ceiling fixture is ideal here. A combination switch and outlet makes provision for use of electric shaver, hair-dryer, vibrator and sun lamp. A night light, switch controlled, adds to the safety of family and guests.

An *Alabax* porcelain vapor-proof shower light, such as that made by Pass and Seymour, is a decided safety factor in the bath. Controlled by a wall switch, inaccessible from the shower, it is a safety factor in any home.

Hallways. Hallways should always be well illuminated, with switches at both ends. Switches with luminous buttons are an added safety factor . . . no fumbling in the dark. When combined with a night light that can be left on when you are out for the evening, you have no fear of stumbling in the dark. If you have an upper and lower hall with a staircase, two-way control of the ceiling lights is a must, and switches should not be placed so close to the stairs that a long reach to the toggle or rocker could cause a misstep. Cellar steps should have the same treatment, and a pilot light on the switch to let you know the light is on will save you cash.

The Workshop. Ceiling lights controlled by wall switches fill the bill here. Install plenty of three-wire grounding outlets for your power tools, *lock* switch controlled to protect youngsters from harm. Place lights where you please, but aim for shadowless light. Glare from gleaming tools can be a nuisance, leading to unnecessary eye-strain, and taking away from the pleasure derived from a hobby.

Laundry or Utility Room. Lots and lots of light are required here, plus outlets for washer, dryer, and in some cases a ventilator. Also, don't forget that you may want to use the ironer while a second load is in the washer. Special purpose outlets are required for automatic washer and dryer . . . check with your dealer and local code before installing these major appliances. If a sewing machine is to be used, provide adequate light for this too. The more outlets the better, just be sure you use the correct type for the appliance.

CHAPTER FOUR

WIRING FOR HEAVY LOADS

WIRING FOR 240 volts and for 120–240-volt combinations varies in detail according to the appliance it is to supply, but certain general points apply to all. The size of the wires, for example, is invariably larger than in ordinary 120-volt house wiring, and is commonly in the No. 6, 8, and 10 range, as current loads are many times greater than in lighting circuits.

A wall oven, for instance, may require close to 5000 watts, a surface cooking burner group, almost 7000 watts, a complete range 12,000 watts, and both 120 and 240 volts. Electric water-heaters use from 1500 to 5000 watts, usually at 240 volts.

Ranges and dryers are often grounded through the neutral (white) wire in the cable, but other requirements in this regard may be specified by the local code. Your only safe bet in 240-volt wiring: check your local code and your local power company for details *before* you begin any 240-volt work.

The important factors start with your service entrance. It *must* be adequate to the load to be imposed. In an old house being modernized, a new service entrance and panel may be needed. If, for example, the service entrance has only two wires instead of three, only 120 volts (not 240) are available. The lighting company can replace it with a 3-wire service, and can provide details on the cost and other factors.

The usual rating of services runs in amperes: 30, 60, 100, 200, and 400. Multiply this by 230 volts for a safe margin and you get wattages of 6900, 13,800, 23,000, 46,000 and 92,000, all of them not necessarily available in all localities.

About Voltages. You may be accustomed to thinking of your house voltages as 110 and 220 because these were the standard figures for many years. But you now hear of 115 volts and 230, 120 and 240, and 125 and 250. Generally speaking you can figure on present-day voltages of 120 and 240, and consider the 125 and 250 printed on many appliances and wiring devices as their particular rating. In general, the difference between 110–220, and 120–240 does not bother appliances. But if you want to be sure of the exact voltage in your area check with your power company. Keep in mind that in all areas periodic

How to install split circuits with interchangeable devices: Diagram below shows a three-wire split circuit for kitchen appliances or workshop. From two circuits and neutral in main switch box, three No. 12 wires are run, and white wire is connected to silver terminals of upper and lower receptacles. A common neutral is used. Red wire is connected to brass-color terminal of top receptacles, and black wire to brass-color terminal of bottom receptacles. This gives two circuits in one box.

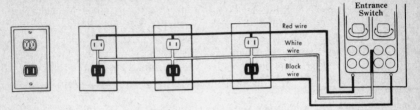

Diagram below shows split receptacles—upper half switch-controlled. In living room, from nearest outlet, white (neutral) wire is run and connected to silver-color terminals of upper and lower receptacles. Black wire is connected to switch and brass-color terminals of lower outlets, and red wire to brass-color terminal of upper outlets.

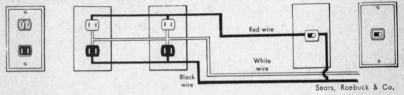

Sears, Roebuck & Co.

variations of a minor nature are unavoidable because of differences in current demand, as when a heat wave results in greatly increased air-conditioner use, or a cold wave keeps oil burners and electric heating systems running.

Split-Circuit Wiring. As outlined in an earlier chapter the three wires that comprise the start of a modern home-wiring system provide both the high and low voltage. Continuations of the same three wires (in the house wiring) may also be used to give extra capacity to 120-volt wiring by providing two circuits on a "split-circuit" basis. The white (neutral-ground) wire serves as one side of each circuit, the two "hot" wires as the other sides. Practically speaking, the white wire connects to the chrome terminal side of two groups of outlet receptacles, providing both with the ground side of a circuit. A wire from the red (hot) wire connects to the brass terminals of one circuit, and another from the black (hot) wire connects to the brass terminals of the other. This provides two circuits with a common (white-ground) wire, permitting the use of a single 3-wire cable for both. Although 240 volts would be available between the two "hot" wires it is not used in the arrangement shown, as the circuit is not designed for the purpose.

Where only the hot red and black wires run to an appliance, as in the case of a water heater, a cable may have to be used that contains only a black and a white wire, but the wiring rules say the white wire should be used only as a ground wire. The answer: paint the ends of the white wire black.

Appliances using only 240 volts but having a ground terminal are often grounded through a third (white) wire in a 3-wire cable. But check your local code and your local power company on this before you do the work.

Interchangeable-Device Receptacles. Split circuit wiring can be done easily with two types of outlet receptacles, both of which fit in standard boxes (switch type). One form of receptacle is the "interchangeable device." This is a single outlet receptacle designed to fit in a locking plate that fastens to the box in place of the usual duplex outlet. The plate commonly has three holes to take either individual outlets, switches, pilot lights, or night lights in any desired combination. It is often referred to as a 3-opening Despard, after Victor Despard, who pioneered it. With it, you can have two outlets plus a night light, or an outlet, with a switch, and a pilot light to remind you that a remote light (as in the basement) is on. Or you can have any other combination.

To wire a pilot light you use a 3-wire cable to lead a branch of the white wire from the light location back to the switch and pilot location. The white wire is connected to one side of the pilot light. A branch from the switch wire leading to the remote light is connected to the other pilot-light terminal. The basic idea is simply to wire the pilot light exactly the same as the remote light. It can be used with a 3-way switch by leading wiring from the remote light to the pilot light. When the remote light goes on the pilot light goes on. That's all there is to it. The pilot light, a neon-glow type, uses so little current as to be almost undetectable on your electric bill.

Duplex Outlet Receptacle. Another type of receptacle is also used in split-circuit wiring. This is the duplex outlet receptacle with a "break-out" section between the paired terminal screws on each side. This break-out is slotted so it can be snapped out with a screwdriver, removing all electrical contact between the two terminal screws on that side. If the break-outs are removed from both sides the two outlets are completely separated electrically and can be wired to entirely different circuits. In the case of split-circuit wiring, however, only the break-out on the brass (black wire) side is removed. That on the chrome side is left intact. Then the incoming white neutral or ground wire can be connected to one chrome terminal and the outgoing white wire to the next outlet connected to the other for a continuous ground. On the brass side one of the two "hot" wires from a fuse cabinet is connected to one of the terminals, the other "hot" wire to the other terminal. This puts each individual outlet of the duplex receptacle on a separate circuit although they have a common ground. In kitchen work areas the result is a much better distribution of small appliance load.

Position the outlets so the upper ones are all on one side of a split circuit, the lower ones on the other, or plan the same result with left and right outlets. Be sure the right ground wire connects to the chrome side all the way through the wiring and the two hot wires (commonly red and black) to their separate terminals on the brass side. *Do not* connect either of the hot wires to the chrome side at any point.

Note that both plugs and outlets are designed for the voltage and amperage they are to carry, and that a plug rated for 250 volts will not fit in a 125-volt outlet and vice versa. This eliminates the danger of plugging a 120-volt ap-

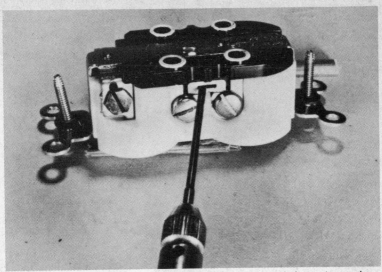

Duplex outlet receptacle for split-circuit wiring has break-out strip which permits severing contact between terminal screws by a twist of screwdriver. Each outlet in the receptacle can then be wired to a different circuit.

pliance into a 240-volt outlet. *Never* use an outlet designed for 120 volts on a 240-volt circuit, or vice versa.

A similar wiring system to the split circuit can also be used with 120 volts, but for a different purpose. Starting from the usual 2-wire black and white wires of the cable, you connect the white wire to the white wire of the 3-wire cable in the usual manner. You then connect the black wire of the 2-wire cable to the black wire of the 3-wire cable *and* to the third wire (always of another color). Run the white wire to the chrome terminals of break-out outlets (separated on the brass side) or to outlets of the interchangeable-device type. You can now insert a switch in one of the hot wires (not the white one) and half of the outlets in the system will be switch controlled, the other half always "live." In a living room or family room this lets you connect all lamps to the switched side, other appliances like fans or TV to the always "live" side. Then, by flipping the switch you can turn out all lights while leaving TV and fans (or air conditioning) in operation.

CHAPTER FIVE

WIRING AN OLD HOUSE

WHEN YOU repair or modernize old wiring there are several important points to keep in mind. First, remember that others may have worked on the wiring before you and before regulations existed in the area. Not all of them were necessarily in the business, and some may have been dangerously unfamiliar with the rules. The older the house the greater the likelihood of the kind of bungled and inexpert wiring electricians often call "home cooking." Color coding of wires and terminals, for example, may have been ignored or followed in some places, not in others. Such conditions are frequently found. You may also find wiring systems like the once common "knob and tube wiring" that are now practically unheard of in residential wiring. Do not, however, decide that wiring must be replaced simply because it is old. If it is in good condition it may last the life of the house.

Checking the Old System. Your best bet is to make a thorough initial check of the electrical system. In general, this is called for if any major wiring blunders are found in the course of a casual inspection.

If, for example, you find black and white wires connected in reverse order at some point in a run of wiring you can suspect other possible errors. Do not correct such errors on the spot. If wire colors are reversed at a connection it is likely that grounding connections are reversed along the run beyond the error point. Correcting the initial error could easily cause a short circuit unless other errors are also corrected. In this type of old work it is more than ever important to use caution. Never attempt even a minor repair until current is definitely off at the main switch. Where improper wiring is a possibility, removing a fuse to deaden the circuit is not enough.

If you find a wiring error, check the entire circuit on which you found it. Do this by shutting off current and removing cover plates and outlet receptacles—not disconnecting them unless errors exist. If errors do exist correct them before replacing the receptacles.

Insulation troubles are a clue to the overall condition of old wiring and are easy to spot during any repair work. If the insulation on wires at the fuse cabinet is cracked or crumbling, there is a fair possibility that the same condition exists along the entire run of wiring, even inside armored cable. To confirm this suspicion, you can check the condition of insulation on wires in outlets along the circuit. If you find the same deteriorated insulation at all points, the wiring should be replaced. The condition can be caused simply by aging of some old forms of insulation or by overheating of the wires, as when undersized wiring is overloaded through the use of an oversized fuse. Such overloading may heat the wires enough to cause a fire within the walls or it may, in mild cases, break down the insulation and create the possibility of a fire later on.

Finding the Path of the Circuit. This is not always easy in an old house but is likely to be simplified by the fact that older wiring generally has fewer outlets. Also, very old installations were often limited to a few circuits.

The usual track-down routine consists of plugging a lamp into outlets throughout the house to determine if they are operative when the current is on. (It's worth knowing which outlets work, anyway.) Then, with one fuse removed at a time, the outlets are rechecked with the lamp. Permanent wall and ceiling fixtures can be checked at the same time by merely switching them on. This wiring rundown also enables you to place a taped-on label on the fuse cabinet cover telling what rooms are supplied by each fuse—very handy when a fuse blows.

Once you know which outlets a circuit supplies it is a simple matter to check the condition of wires along it. Any circuit with badly damaged insulation requires new wiring. If the wiring was installed when the house was under construction it will be stapled to the structure and cannot be removed.

In very old houses (originally without electricity), however, the wiring may have been simply "snaked" through walls In this event it may be possible to fasten new cable to the old cable and pull the new cable through as the old cable is pulled out. Where the old cable cannot be removed, it can be pushed aside to make room for the new cable at boxes. Exact methods depend on the situation. If complications exist it is often easiest to eliminate them by making hand-sized holes in the wall where new cable must be led to boxes and old cable removed.

Where walls are to be preserved without marring multi-outlet raceway (see chapter on Surface Wiring) is often the ideal answer. This and its non-metallic relatives are designed to run along the wall surface like a molding or baseboard (or atop an existing baseboard) with little or no need for breaking into the wall.

Knob and Tube Wiring. This is an old wiring system still found in thousands of homes and buildings, though it is now a part of the electrical past. It gets its name from the fact that porcelain knobs were used to anchor the wires to the building framework, and porcelain tubes were inserted in bored holes to serve as insulating bushings where the wires passed through. As the wires were well insulated, many of these systems are still functioning and in serviceable condition. In modernizing, however, it is not wise to consider merging new wiring with this form. Instead, start any expansion at the fuse panel and run a modern system to all new outlets.

In checking over the wiring of a very old house keep in mind that many of the basic safety requirements established by today's National Electrical Code were unknown to many in the early days of home electrical power. This makes it essential to check on the grounding provision of any house in the really antique category. The wiring may have been installed before the house had a piped water system, and may, therefore, be grounded to a driven rod. If other aspects of the wiring are correct, it is a simple matter to ground it to the water pipe. If the water pipe has less than 10' of underground length both a rod and pipe grounding should be used.

In all old wiring look for the hazards created by exposed cable splices (not enclosed in approved boxes) and sometimes found in the faulty wiring jobs. If you spot such a splice in a run of armored or nonmetallic cable it may not always be possible to remake the splice inside a box because of limited cable length. In this event it is always possible to use two junction boxes with sufficient length of new cable between them to allow for in-the-box splicing, as required by the Code. Naturally, make all such repairs with current *off*.

CHAPTER SIX

SURFACE WIRING

SURFACE WIRING is the oldest form of wiring. It is mounted on the surface of the walls inside a house or building, and was the most common procedure in the early days of electric power. It is still the simplest means of bringing wiring up to date in an old house. It minimizes the need for cutting into walls and permits future wiring expansion or changes with very little work, as compared to in-the-wall wiring. Used in new construction, it provides a far greater number of outlets for a given amount of work.

Materials necessary to add a single duplex outlet with ordinary armored cable. Cable must be snaked through wall and box mounted by cutting a hole in the wall.

Several forms of surface wiring are available for simple additions to existing wiring. For major expansion or overall wiring, as in new construction, the metallic raceway form and related multi-outlet plug-in strips such as the familiar Wiremold, are the most versatile, and are UL approved. They are also approved by most local codes.

The different types are described in detail in the paragraphs that follow. As shown by the photos, they are among the simplest wiring forms to install. The nonmetallic types are often the easiest to install on small expansion wiring jobs. The metallic raceway and related types offer the widest range of use, including complete wiring systems. And some makes, such as Wiremold, offer fittings that permit connection with any other approved wiring system.

The parts you need for a surface wiring job are available in hardware and electrical supply stores. You can expect to pay about a dollar per running foot for materials. This is more than the average materials cost for conventional in-the-wall wiring. But if you figure in the extra advantages, plus the saving in labor by doing your own work, you can put in surface wiring for about the same cost as ordinary wiring installed by a licensed electrician.

Another advantage is that when rewiring an old home, cherished panelling or beautifully papered walls need not be cut into except for one feed-in for each run of raceway. Feed-in can be traditional BX cable, as most raceway has fittings to adapt to BX. The only opening will be concealed by the cover plate when the job is completed.

Plug-in Strips. These are of rigid plastic, each 12″ long. The starting section plugs into any outlet; each succeeding section plugs into the end of the one ahead of it.

Where you wish to have outlets, you insert a section that has receptacles molded into the strip. In between, you use plain spacer strips to carry current.

This type of wiring is called "nonmetallic surface extension." Its heavy-duty No. 12 wires will safely handle a full circuit load; an ordinary extension cord won't.

Multi-outlet Strip. A similar type of surface extension, this strip has no fixed outlets built into it. It is a flat, ribbon-like flexible plastic strip with No. 12 wire embedded under a lip at each edge. You put up the strip first, then clip separate receptacles into it wherever you want them. Prongs on the clip contact the wires. You buy only the outlets you need. Later you can add more or rearrange the ones you have.

This type of strip does not plug into the wall like the rigid type. The starting end is wired permanently into the existing wall box.

Both types of plastic wiring look best if kept low—either on or just above the baseboard. Run vertically up a wall, they may become conspicuous.

Metallic Multi-outlet Raceway. This resembles the old raceway wiring. It is a little more expensive and harder to install than the other new surface types. But it has a number of special features. The wires run inside a slim metal channel capped by a snap-in cover plate. Outlets, switches, and light fixtures can be mounted right in the channel. If you want to change them later, you just pry off the cover plate to expose the wires.

Nonmetallic surface extension can start from an existing outlet simply by plugging it in like an extension cord. Outlet receptacles may be located where needed. Approved types can be fastened to the baseboard with screws provided for the purpose.

You can also add extra circuits depending on the type of raceway you buy. Several sizes are available, some holding up to five or more separate two-wire circuits. The biggest channels are expensive and you probably won't need them, but it's a good idea to allow for at least one future circuit.

INSTALLING SURFACE WIRING. You can use metal raceway as a baseboard itself, eliminating wood trim. If it touches the floor, however, you'll need a special type—somewhat costlier—that raises the outlets to the required legal distance from the floor. Check the code for the brand you use. You can also get a rounded metal cap strip that clips to the top of the raceway. This gives the neat appearance of a quarter-round molding and can also be used to house low-voltage wiring for doorbells, hi-fi systems, and telephones.

Raceway is wired by making a permanent connection at one end to an existing wall box. The outlets come prewired in long strings, eliminating connections between the receptacles. You can buy them spaced from 3″ to more than 5′ apart. In addition to standard outlets, you can buy double receptacles wired so that one half is always on, while the other half can be controlled from a switch. You can also get the newer three-way grounded receptacles recommended for power tools and appliances.

Where to Use the Strips. The nonmetallic types (both rigid and flexible plastic) can be run only in *exposed, dry areas*. You cannot use them in basements, attics, garages, or other possibly damp locations, and you cannot run them through or inside a wall. They must not be allowed to touch the floor or come any closer than 2″ above the floor. In mounting them, attach them only to wood or plaster surfaces, away from all metal except for the one connection at the wall box.

Metal raceway is the only type of outlet strip that can be used in basements, attics, or other unfinished areas. Because of its metal channel, it provides a continuous ground connection back to the fuse box. The other, nonmetallic strips do not. Even raceway, however, should be kept away from any excessively damp areas, such as wet cellar floors or walls. It should also be fitted with the three-way grounded type of receptacles now required by the National Code.

None of the strips should be placed near excessively hot locations, either. If you have baseboard heaters, don't mount the strips directly against them. Fasten a 1″ by 2″ wood spacer-strip above the heaters, flat side against the wall, then run the wires above the wood.

How Many Outlets for Your Needs? There's no legal limit to the number of outlets you can add so long as each strip stays in the same room where it starts. You cannot run a strip from one room into another room through a wall.

There's no point to adding more outlets than you can conveniently use—they're one of the most expensive parts of the deal and will just run up the cost. For baseboard use, an outlet every 30″ to 60″ is adequate. For a kitchen counter or workbench, you may want one every 12″ or less. It's also wise to remember that adding outlets does *not* increase the current-handling capacity of the original branch circuit. You must take care not to turn on any more lights or appliances at the same time than you normally would.

In some homes, the outlets in each room may be fed by two circuits instead of one. If you tap into only one existing wall box, you are thus likely to waste one circuit while overloading the other. In this case, it's best to split your wiring, running two strips from separate wall boxes, one on each circuit.

The plastic strips do not provide for switch controls, but there's a trick you can frequently use to get a switched circuit. Often one outlet in each room is wired to a wall switch. If you connect the strip to this box, all of your new outlets will also be controlled by the same switch.

To start multi-outlet system with metal raceway, simply lead cable into channel through knockout in back section, which will later be screwed to wall. Front cover of channel has not been snapped in place. Current, of course, is off.

Wires from cable are spliced to those in multi-outlet receptacles which are mounted in outlet openings in cover of metal channel. Distance between openings may be 30″ or 60″, or they may be in clusters for work areas, as in a shop.

Outlets are held firmly in cover openings of channel by spring clips on rear of cover, which is lying face down.

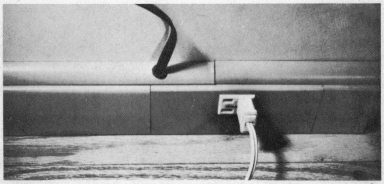

Completed section of multi-outlet system. Snap-on strip may be added on top of regular channel to carry aerial or hi-fi wiring. Outlet here is contained in separate section of cover. You can place a single outlet or a cluster anywhere you want.

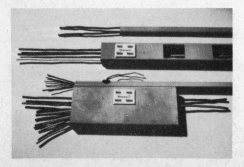

Multi-outlet raceway comes in various sizes (from top): small single-circuit raceway to run around doorways, windows; 3-wire raceway which permits switching one side of each duplex outlet; baseboard raceway with multiple circuits plus outlets.

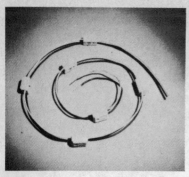

Pre-wired receptacles for multi-outlet system are spaced along wires to match cover openings in channel cover sections. Nothing to do but snap them into holes, snap on cover, after making just two wire splices to start run.

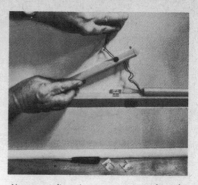

Here, medium-size raceway multi-outlet channel is mounted atop existing baseboard. Top strip, containing doorbell wiring, is being snapped in place.

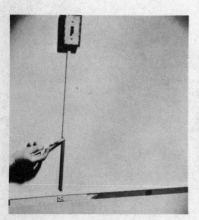

All types of raceway and multi-outlet channel made by Wiremold can be interconnected. Here, small 2-wire raceway leads from multi-outlet run up to wall switch.

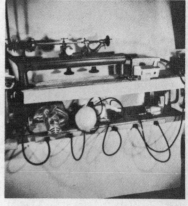

Used on workbench, multi-outlet system permits tools to be kept plugged in. If more than one tool is likely to be used at one time, use larger channel and include two or more circuits.

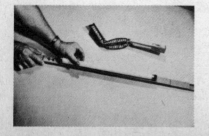

Fluorescent lighting fixtures are made to fit the channels and can be snapped in like cover section. Separate or self-contained switches are available. Fixtures are handy in close quarters, as above counters, below upper kitchen cabinets.

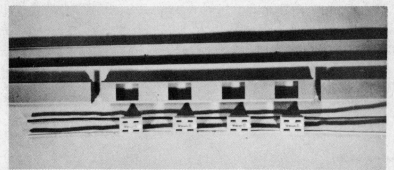

Outlet clusters like this are available in wide variety of spacings with cover sections to match. They can be inserted wherever needed in run of system.

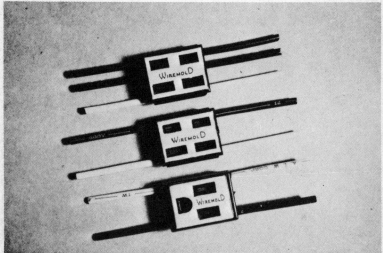

Three types of outlets: Top, 3-wire outlet allows one side of each duplex outlet to be controlled by a switch while the other side remains always live. Center is unswitched 2-wire outlet, always live on both sides of duplex. Bottom is grounded single-outlet receptacle, now standard, which fits all 3-wire shop tool cords.

Nonmetallic surface system uses hard plastic fittings for corners, elbows, and T's, also for start of run from existing outlet box. Run of wiring is flexible plastic strip. Outlets may be placed in the strip wherever needed and locked in place by giving them a quarter turn or using locking lever. Outlet can be moved at any time by reversing the procedure. Strip is fastened to baseboard or wall with nails or screws provided for the purpose.

How strip and fittings are joined. Unit with wires protruding is special cover plate for starting wiring run from existing outlet box. Fitting at lower right starts run from BX cable.

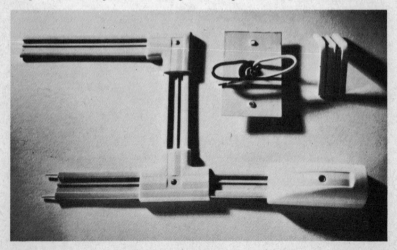

Strip's plastic is trimmed at end so wires can be inserted in starting cover plate and connected. Fitting is adjustable to angle.

Once strip is mounted on wall, outlets can be inserted wherever needed. Tip them to angle like this, then turn to level position to lock them in place.

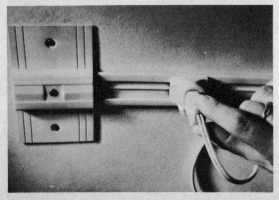

Here, outlet receptacle has been locked in place and cord is being inserted in it. The receptacle can be easily removed and relocated at any time.

How metal and nonmetallic channels go around corners. Metal channels use corner fittings; nonmetallic channels simply bend.

Sectional, or link, surface wiring is another nonmetallic form. It starts by simply plugging in first section at a regular outlet. Flexible section at top takes it around corners. Blank section below it is used for runs where no outlet is required. Center section contains three outlet receptacles, can be inserted anywhere along run, or used in groups. Plug-in starting sections are made to start from either vertical or horizontal outlets. All sections are 1' long, fastened to wall with screws provided for the purpose.

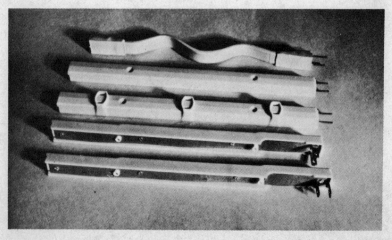

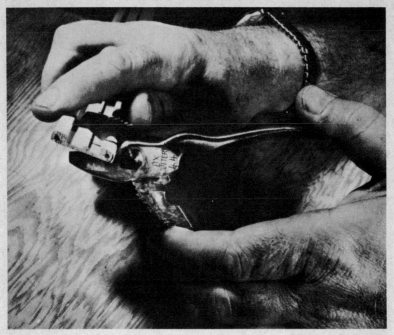

Special strip-baring pliers speed the job of trimming ends of flexible-strip nonmetallic surface wiring. Jaws cut through plastic without cutting wires embedded in it.

CHAPTER SEVEN

OUTLET REPAIRS AND INSTALLATIONS

OUTLET REPAIRS. The commonest cause of outlet trouble is poor electrical contact. This is likely to develop even in a high-quality receptacle after long use. It becomes apparent when appliance plugs fit into a receptacle loosely or fall out. Do not try to overcome it by bending the prongs of the plug on the extension cord, as the bent-prong plug may damage a good receptacle if it is inserted into it.

Another, though less frequent, cause of trouble is poor contact between the wires and the terminal screws of the outlet. This can sometimes be detected through the noise it causes on a radio when an appliance places an electrical load on the affected outlet. If you hear a continuous sputter when an appliance is plugged into one particular outlet, your radio may be giving you an important clue to trouble.

Try several appliances in the outlet. If the sputter persists with them all but ceases when they are unplugged, you have ample reason to suspect the outlet. If the appliances cause no sputter in other outlets on the same circuit your first check points are the terminal screws on the outlet.

Shut off the Current. This can be done by turning off the main switch (stopping current flow throughout the house), or by removing the fuse that supplies the circuit on which the suspected outlet is connected. Be absolutely sure, however, that the current to that outlet is off. You can use a test light with its prongs in contact with both terminals of the outlet. But if you have any doubt pull the main switch. If you have an accurate chart of your individual circuits, however, the fuse-removal procedure lets you work on one part of the wiring without shutting off current to the entire house, sometimes a considerable inconvenience.

Tighten Terminal Screws. With current definitely off, remove the cover plate from the outlet by taking out the single screw in its center. Then remove the two screws (one at each end) that hold the receptacle in the box. Place all screws in a saucer or ashtray, as they are easily lost if not placed in a safe container.

Next, pull the receptacle straight out of the box far enough to expose the terminal screws. Try the attached wires with your fingers. If they move easily under the heads of the terminal screws you may have found the trouble. Just one loose screw is enough to cause the trouble. Be sure that the wires are looped in the direction that the screws turn to tighten. You can do this by loosening the screws and examining the wire ends.

If all's well, retighten the screws, place the receptacle back in the box, and replace the screws that hold it. Then replace the cover and try an appliance with your radio plugged into the same circuit. The current, of course, must be turned on for the test. If the sputter is gone you have found the trouble.

Replace Receptacle. If tightened terminal screws don't cure the trouble, the receptacle itself may be the cause. Shut off the current, follow the steps just outlined, and remove the receptacle, disconnecting the wires instead of tightening the terminals. Replace it with a new one, connecting the black wire to the brass terminal, the white to the chrome. If wires are attached to both pairs of terminal screws be sure all are tightly connected. In most cases the new outlet will cure the trouble.

If it does not, and plugging in an appliance results in a lasting noise, your trouble may lie in an outlet preceding the one in question along the same run of wiring. So, check the others in the same manner. While this type of trouble is not too common, it is often an annoying puzzle to those unfamiliar with it. Also, a loose connection can be dangerous, as it causes arcing that may prove a fire hazard.

Do not attempt to repair an outlet receptacle. The contacting strips are seldom accessible and even if they are should not be rebent to improve contact. When the pressure of the spring stock fails after long use rebending is very likely to result in breakage. The only safe answer: when an outlet receptacle fails, replace it.

ADDING NEW OUTLETS.

If your house is an old one with too few outlets, perhaps with the only ones in some rooms located where furniture must be placed, the problem is not a difficult one to solve.

With current to the outlets shut off, remove the cover plate and the outlet from the box, leaving the wires connected. If only two wires are connected to the outlet, it is a simple matter to extend the wiring to another one.

If all four terminal screws (two chrome, two brass) are connected to wires, examine the other outlets in the room in the same manner. In many old homes there are only two outlets in bedrooms, both on the same circuit, and one at the end of the run. The end one will have only two of its four terminal screws connected to wires. It is easiest to run additional wiring from this one, using the two vacant screws to start it.

Locating the Outlet. Begin by marking the locations for the new outlets on either side of the one from which the wiring is to start.

With the current off, open the starting outlet and disconnect the receptacle from the wires. Remove the baseboard along the side of the room under the outlet. Make an opening in the plaster or wallboard in the area from which the baseboard was removed, directly below the outlet. Using a screwdriver and hammer, tap open a knockout in the bottom of the outlet box.

Cut a channel into the plaster from the opening directly under the box to a point directly under the location of the new outlet to be installed. The channel in the plaster must be large enough to take the type of cable (armored or otherwise) when the baseboard is replaced. Make an opening through the plaster or wallboard below the new outlet and in the area to be covered by the baseboard.

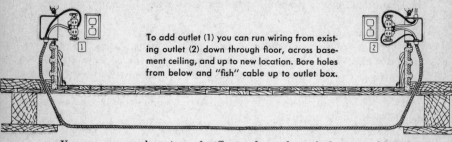

To add outlet (1) you can run wiring from existing outlet (2) down through floor, across basement ceiling, and up to new location. Bore holes from below and "fish" cable up to outlet box.

You can now push a piece of stiff wire down through the opened knockout in the old box to the opening in the wall below it. Reach through the hole with fingers or pliers to pull the lower end of the wire out. Then bend the ends a little so the wire won't slip out altogether. It will be used to pull cable into the box.

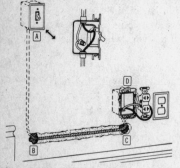

If new outlet is to be switch-controlled, cable is led down from switch box (A), then led along channel (B and C) chiseled in wall (with baseboard removed) and up to outlet (D). When baseboard is replaced, cable is concealed.

Stringing the Cable. Next, string the cable you plan to use along the wall from the old outlet down to the hole below it, along the horizontal run, and up to the new outlet location. This is all done along the wall surface. No stringing through holes is necessary at this point—you are simply gauging the length of cable you will need for the job. You can use cellulose tape to hold the cable in position if you haven't a helper. Mark it at each end, allowing for 8″ of free wire in each box. Then cut the cable as described in an earlier chapter, removing enough armor (from armored cable) to provide the needed free wire at the ends.

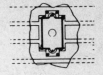

When cutting hole in wall for new outlet or switch box, cut through one lath completely, notch lath above and below it. This is more rigid than when two laths are cut away completely.

Connect the incoming and new outgoing wires to the starting receptacle and lead the new wiring to the hole below the new outlet location. The new outlet box may be mounted in the wall with a toggle-like clamp arrangement made for the purpose or with other ready-made devices. All are designed to grip the plaster wall or wallboard so you need not worry about fastening the new box to the wood framing. The clamp-type system is usually easiest to install.

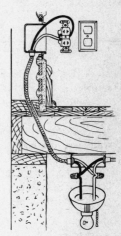

Left: To connect new first-floor outlet to wiring on basement ceiling, drill upward for cable so hole emerges inside wall where outlet will be.

Right: Lead cable up to new outlet, using fish tape. Connections are always black to black, white to white, as shown.

Sears, Roebuck & Co.

Electrical Connections. Once the new box is in place you are ready for your electrical connections. It is a good idea to open the knockout for the incoming wiring to the new box before it is mounted in the wall. If further wiring is to be led from it open the knockout for that, too. This way, the impact of opening the knockout is not applied to the wall and the cut wall edges around the box.

To pull the cable up into connecting position at the new box, first place the fiber bushing in it and attach the connector by tightening its setscrew. Then make hairpin bends in the wire ends and hook the end of the stiff lead-in, or snaking wire, to them. A turn or two of friction tape over the joint will prevent it from unhooking. Now you can pull the cable up to the new box. (This is the same system used in leading the cable to the starting box.) Once the connector has been eased through the knockout hole, the locknut may be spun on and tightened with a few hammer taps on a screwdriver, as outlined previously. To complete the job the wires are connected to the new receptacle,

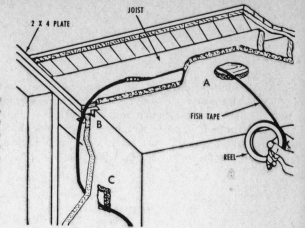

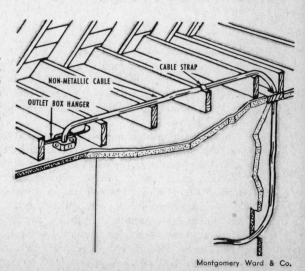

To lead new wiring from ceiling (A) to wall (C), it is necessary to remove small section of plaster at B where wall and ceiling meet so 2-by-4 plate can be notched to pass cable. After cable has been led through, plaster can be patched to conceal hole.

black to brass screw, white to chrome screw. The remaining two terminal screws may be used to extend the wiring to other outlets in the same manner.

Supporting the Cable. Along horizontal runs behind baseboard, the cable should be properly supported. This can be managed by stapling armored cable, using staples long enough to get a bite in wood behind the plaster or wallboard. With nonmetallic sheathed cable, the usual strap supports should be used with screws long enough for the purpose. Never force the baseboard against the wiring behind it. If a staple or other fastening (or the cable itself) holds the baseboard out from the wall at any point, use a chisel or other appropriate tool to make a recess in the back of the baseboard so all parts fit

If attic is unfinished, wall-to-ceiling wiring is easier. Hole can be bored from attic into wall; outlet box and hanger are easily mounted from above. Use plywood panel for working platform in unfloored attic. Nonmetallic cable is shown here, but BX or conduit may be used, depending on local requirements.

Montgomery Ward & Co.

easily without crushing pressure or chafing. As this is not a common problem in this type of work it is mentioned only so that it will be handled correctly if it occurs.

Overloads. A large number of outlets on a circuit does not overload it. But a large number of appliances plugged into those outlets can blow a fuse if they exceed the capacity of the circuit. For this reason keep in mind that adding outlets to a circuit is merely a means of increasing *convenience*. You do not have to crawl behind furniture to find an outlet. And you do not need a maze of extension cords to supply the lamps you need.

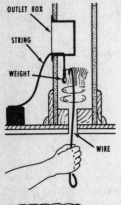

Here is a trick for fishing wire through holes. Let weighted loop of string dangle through cable knockout in box. Hook it from basement with coat hanger wire, then use string to lead fish tape or cable through.

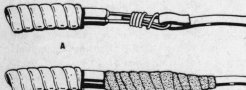

Attach fish tape to wires of cable as at A. Wind with plastic tape, as at B, to make joint smooth and prevent it from catching on corners or hole edges.

Montgomery Ward & Co.

Remember, however, that the overall capacity of the circuit remains the same when you add outlets. If the fuse is rated at 15 amperes, it will still blow when that limit is reached—and it is actually easier to reach with additional outlets. If you need greater capacity, freedom to use more lamps or appliances, you need more circuits, possibly of higher capacity. This calls for new runs of wiring from the entrance panel to the area of use. In some cases, as in old 2-wire entrances, it may call for heavier entrance wiring and larger service panels. Check this type of requirement with your local power company. For example, if the lamps and appliances you normally use in a room blow fuses frequently, you have a circuit that cannot take the load. The only way the problem can be solved: add another circuit—or split the load with an existing circuit. The latter solution is not always easy in old houses, as circuits were often laid out floor by floor, a plan now recognized as impractical. If a fuse blows, the whole floor is affected. A better and more modern method is to use at least two circuits in major areas and rooms. If a fuse blows on one circuit, you have another still working.

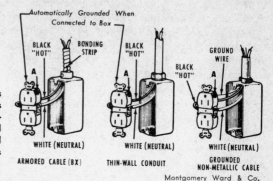

Three ways that a continuous ground is provided to boxes throughout the system. Armored cable and thin-wall conduit automatically ground boxes; nonmetallic cable has separate ground wire.

Montgomery Ward & Co.

Grounding Outlets. For the first time, in 1962, the National Electrical Code required that only grounding outlets be used. Previously it required only that this type of outlet be used in certain rooms where other types might create a hazard.

Basically, the grounding outlet is designed to assure that the parts of an appliance will not be dangerous to touch. If they are grounded, they are on the same side of the circuit as your plumbing, for example, and therefore should be safe to touch. (Touching some older appliances and plumbing simultaneously can result in an electrical shock.) It should be noted, too, that some of the newest appliances, particularly certain power tools, are made with a complete outer housing of insulating material with insulating sections in shafts so that all exposed parts are electrically "dead."

To understand the development of the modern grounding receptacle and cord systems, it is necessary to realize that nongrounding systems were developed in the early days of house wiring. You have probably heard that if you stand in a filled bathtub and turn on an electric light you can be electrocuted, and in many instances you can. If you touch a metal part of an electrical fixture that happens to be on the "hot" side of the circuit while you are standing in a filled tub that is on the "grounded" side, you will be the object through which the current flows—and the results can range from extremely unpleasant to fatal.

The 3-prong plug-receptacle system assures that what you touch will be on the same side of the circuit as anything else you are likely to touch. In time, properly used, this will eliminate many of the old-time hazards of household wiring. But never count on it without testing. A connection may be reversed. That's why inspections are made, and even inspectors can make mistakes.

When you do a wiring job, take advantage of the grounding outlets. An ever increasing number of modern appliances are being made with grounding plugs to match them, and you can use standard 2-prong plugs in them, too. The usual brass and chrome terminal screws are used as usual plus a single, green terminal screw. This one is connected to the third (U-shaped) hole in the receptacle and also to the metal tabs at the ends that are used to fasten the receptacle to the box. Thus, the third prong of a grounded plug is grounded to the box—if the receptacle's fastening tabs are in good contact with the box. To make sure of a thorough ground, a wire is run from the green terminal screw to a screw in the box.

Four kitchen appliances can be operated at once on appliance unit without risk of blown fuses. Two 20-amp. circuits usually feed this type, with a pair of outlets on each. Wiring must be No. 12 or larger.

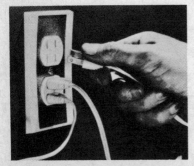

Four-plug outlet lets you connect twice as many plugs as usual but still fits into standard box that provides only two outlets with ordinary receptacles.

Clock outlet supports electric clock on hook at top, provides recess for plug and short cord behind clock.

General Electric Co.

Floor outlet has two threaded covers. One is solid to keep out dirt when outlet is not in use; the other has hole to admit cord when outlet is in use.

Pass & Seymour, Inc.

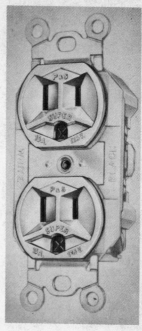

Dust- and moisture-resistant receptacle has neoprene gasket behind openings. Slits in neoprene at openings admit prongs from plug but close again as soon as plug is removed.

Pass & Seymour, Inc.

Combination outlet receptacle, switch, and pilot light fits standard switch box, provides reminder when remote light or appliance is turned on.

General Electric Co.

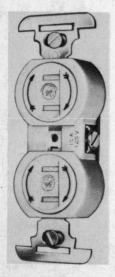

Safety outlet has plastic disks which must be turned to open slots in receptacle before plug can be inserted. Turning requires adult strength in fingers.

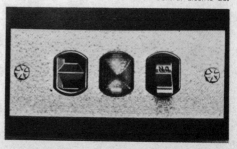

If you are using nonmetallic sheathed cable, you must use the type that contains a bare grounding wire in addition to the two regular wires. At each box the grounding wires of all cables entering must be connected to each other, to the green terminal screw, and to the box. This assures a continuous ground.

Appliances and tools designed for 3-prong grounding plugs (and 3-wire cords) have their outer body or shell grounded through the cord and its contact with the ground of the wiring system. As the 3-prong plug can be inserted in only one position, the "hot" and "ground" sides of the circuit cannot be reversed by turning the plug around, as can happen with ordinary 2-prong plugs. The grounding system protects you when you must use a tool while standing on a damp floor, for example; but do not assume it will protect you from misuse of an appliance, as while standing in a tub. The other side of the circuit is still present inside the tool or appliance.

Special Outlets. For safety when small children are about, you can buy outlets with spring-loaded covers to prevent objects other than plugs from being inserted in the openings. Another type has an internal mechanism for the same purpose. For a wall clock you can buy a clock-hanger receptacle that supports the clock and permits the use of a short cord, completely concealed behind the clock. There are also combination switch-and-outlet receptacles, and others that contain their own fuses. Each of these can solve a special problem for you.

General Electric Co.

Hinged covers on outlet prevent dirt from entering slots. This is 3-prong receptacle for use with heavy appliances and power tools.

CHAPTER EIGHT

PLUGS AND CORDS

PLUGS. Wall plugs are available in many types. Some are an integral part of the wire from the appliance, molded to the cord. Others are attached with terminal screws to the wires. Some have handy shapes to facilitate removal from a receptacle where frequent changes of appliances are made at the same outlet. Modern design has developed other forms to allow a cord to run out of the side of the plug, parallel to the wall, rendering the plug and wire almost invisible, a desirable feature in living areas where plugs and cords might be unsightly, serviceable though they be.

Wall plugs that take wire without stripping are the easiest for the amateur to use in replacement jobs, and no tools are required. Simply open the locking lever on the plug, insert the wires, unstripped, close the lever firmly so that the prongs in the plug make a good contact, and the new plug is ready for use.

When a plug no longer functions, the safest thing to do is replace it. Bending prongs, wiggling for a better contact, are makeshift at best, and replacement is the safest and, in the end, the cheapest answer.

Replacing Molded Plugs. When a molded plug fails, the only answer is to cut the cord and replace the plug with one suited to the receptacle where it is most used. For instance, if old wiring has been replaced with raceway, the duplex outlets are very close together, so slim plugs are called for. Slim plugs fit side by side in the duplex outlet, whereas old-fashioned round plugs conceal one side by bulk alone. The slim plugs may be the type that requires no stripping, as described before, or the conventional type with terminal screws. In any case, if the plug has to be removed often, look for a design with a handy grip for easy removal. Outlets in raceway hold a good grip on plugs, thus are safer than most where small fry may be tempted to remove a cord.

Plugs with Terminal Screws. With the conventional type of plug with terminal screws, before discarding the old plug, check to see if the terminal screws are snug. Sometimes mere tightening of the screws can save the day. However, if there is any sign of fissure in the plug itself, replacement is the answer.

Plugs and plug-in outlets (top row, l. to r.): Flat 2-wire plug for use on closely spaced outlet receptacles, multi-outlet raceway; round plug with grip-neck, allows room for Underwriter's knot; quick-connect plug with slide-in top; 3-wire adapter for 3-wire plug in nongrounded outlet; series-wired plug. Bottom row (l. to r.): Nite-lite adapter plug; standard bulb plug; plug-in nite lite; circuit-breaker plug; multiple-outlet plug.

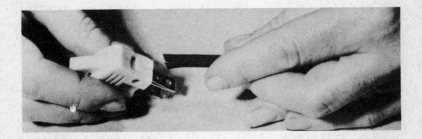

Lever-operated cord plug merely requires inserting end of standard lamp cord in opening when lever is up (above). By pressing lever down (below), pins penetrate cord, making electrical contact with each wire and locking cord in plug.

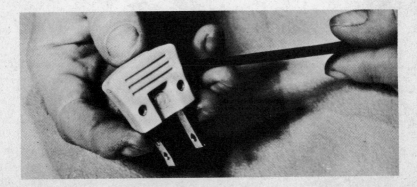

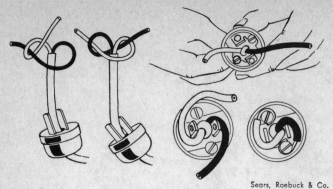

Sears, Roebuck & Co.

Underwriter's knot prevents wires from slipping off terminal screws when cord is accidentally pulled. Loop cords as shown, then pull down gently into recess between prongs.

If by chance the underwriter's knot was used on the original, don't cut the plug off; rather, loosen the screws and gently work the worn plug off the wires. In this way you can duplicate the knot when replacing the plug, as the bends in the wire will be a sufficient guide for redoing the knot. The underwriters' knot is a safeguard if there is someone in your household who is a cord-jerker.

When replacing the wires in a new plug, always give the fine wires a firm twist before attaching to the terminal screws, and wrap the wires around the screws in the direction the screws turn. This makes for a better contact. Also snip off any stray strands that might short across the plug, causing sparks, or blowing a fuse.

To attach cord to screw-terminal plug, run wires around prongs, loop bare ends around terminal screws, and tighten screws. If there isn't room for Underwriter's knot, use small wrapping of tape for same purpose.

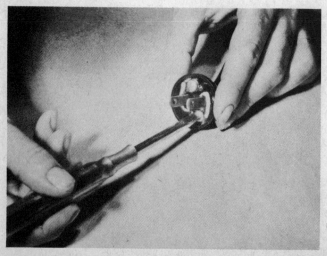

Baring Wires. When the job calls for baring the wires, as when a molded plug is cut off, separate the two wires very carefully with a single-edge razor blade or a trimming knife, taking care to cut in such a way that the wires do not lose their individual insulation. For baring the wires, a penknife is easier and safer to use—a razor might cut some of the fine strands. A gentle scraping will also remove any corrosion from old wires, assuring a better contact with the new plug.

In the event that a three-prong grounded plug has to be replaced, be very sure that the two wires going to the regular prongs are not reversed, and that the third wire is connected to the prong the original was connected to. Duplicate the original connections *exactly.*

Twist-Lock Plugs. Where a rugged, dependable locking connection is necessary, as between portable electric tools and the power supply, "Twist-Lock" plugs, manufactured by Harvey Hubbell, Inc., fill the bill. The locking action is simple: plug in, twist, and it's locked. Safe positive contact is assured, without the nuisance of frequent pull-outs, so often encountered when using portable tools inside or out of the house. For electric hedge trimmers or lawn mowers, these plugs are indeed a boon. Slightly higher in cost than ordinary plugs, they soon earn their keep in convenience and time saved. Special types are also available to the boat buff who takes his power tools to dockside.

Circuit-breaker plug fits in outlet receptacle, accepts plugs from tool or appliance cords. If overload occurs, button pops up and tool or appliance is automatically disconnected.

Circuit-breaker cord plug can be used to replace ordinary plug on appliance cord. It automatically disconnects appliance or tool in case of overload, can be reset by lever.

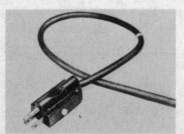

Hopax Electric, Inc.

Polarized circuit-breaker plug does same job as others but is designed for tools and appliances with 3-prong grounded plugs. It is reset by button on side.

Heavy-duty plug for tool and appliance cords (left) has setscrew to hold it to outlet, to prevent accidental pull-out during use.

Heavy duty extension cord with four grounding outlets for power tools (above) can be hung in workshop. Just plug in cord, and multi-outlet unit is ready for work.

Series plug (right) is inserted in outlet, and lamp cords are plugged in two pairs of slots. Lamps then are connected in series and use less current.

ELECTRIC CORDS. The types of cords for household use are varied. Some appliances, such as flatirons and heaters, require cords specially insulated to withstand heat. If a cord is to be connected and disconnected frequently, look for a very flexible type that meets the requirements of your appliance. Some of the plastic-covered cords tend to stiffen and become difficult to handle at low winter temperatures, as in unheated garages or tool sheds. For this type of situation favor a rubber-covered cord.

Very old cords, and those that have been subjected to considerable heat near radiators and hot-air ducts, should be checked frequently for signs of cracking. If cracks are present, flexing may crumble away the insulation, leaving bare sections of wire and a serious hazard. The remedy: replacement.

In all cords, each conductor is made up of many strands of fine wire to provide flexibility. A cotton covering over the twisted strands retains full flexibility by keeping the insulation from bonding directly to the strands. The type of insulation and its thickness varies according to how the cord is to be used.

Underwriters' Cord Type SPT. Commonly used for lamps, radios, and other small appliances, this cord has insulated wires embedded in plastic. Made in colors, it is durable and inexpensive. Type SP is similar but insulated with rubber instead of plastic. Use it for low-temperature flexibility. Both are commonly available in No. 18 and No. 16.

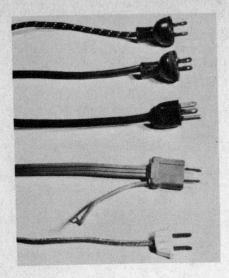

Five types of electrical cord (from top): Asbestos-insulated cord, used for irons, toasters, etc.; 2-wire power-tool cord; 3-wire power-tool cord; heavy-duty 3-wire cord, often used for air-conditioners; common lamp cord.

Underwriters' Type S. In this cord, each wire is rubber-insulated. The two wires are made into a round assembly with paper twine or jute used to fill out the form. The tough outer skin is of high-quality rubber. This type of cord can take plenty of punishment. Another rubber-covered type is SJ, but the outer skin is considerably thinner than that used in Type S.

When neoprene is used in place of rubber, you gain the added advantage of oil-resistance, and the Code designations become SO and SJO rather than S and SJ.

Heater Cord. When a cord has to tolerate considerable amounts of heat, as in irons, toasters and other heating devices, "heater cord" is required. Code Type HPD fills the bill here. A layer of asbestos covers each wire completely before twisting. A cover of rayon or cotton wraps it all. Cotton is considered to be the more durable covering.

When replacing any cord, be sure to duplicate exactly the kind used by the manufacturer. Don't assume that a similar-looking cord will do the job. When in doubt, take a sample of the old cord to your electrical supply house or hardware store to make a perfect match, both in wire size, insulation, and the outer cover. Also match the length of the cord as nearly as possible. An overly long cord may affect performance because of increased resistance.

Prohibitions Against Flexible Cord. There are instances where flexible cord cannot be used. The Code prohibits its use as a substitute for fixed wiring in a structure, where run through holes in walls, ceiling or floors, where run through doorways, windows, or similar openings, where attached to building surfaces (although insulated staples for this purpose are sold in hardware stores and people persist in using them), where concealed behind building walls, ceilings, or floors.

Flexible cord shall be used only in continuous lengths without splice or tap, according to the Code. However, in an emergency many a splice has been made to tide over a Sunday or holiday, but take every precaution if this necessity arises.

Repairing Damaged Cords. When repairing a cord, never thoughtlessly cut the cord without disconnecting it. The least damage would be a blown fuse, but more serious consequences could result.

To repair a cord in an emergency, cut away the damaged section, and separate the wires, then cut the two wires so the splices will be staggered. This method makes for a smoother, safer joint as there is less chance of shorting across the wires in case of insulation failure, which could happen if you were short of insulating tape. Make a pigtail splice in each wire, making sure you follow the original twist, else the wires will tend to unwind. Tape each wire splice thoroughly with plastic electrical tape, then wrap the entire dual splice with plenty of tape. At the first opportunity, replace the entire cord for the safety of your home and family.

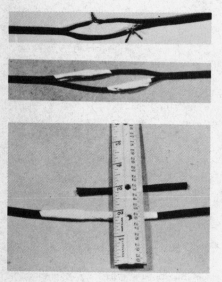

Code does not approve cord splices, but they are often made in emergency. Top: Clip wires so splices are not directly opposite each other. Separate wire strands into two or more bunches and twist each bunch together. Then twist splices together and bend them back against wire in opposite directions. Center: Tape finished splices separately, then tightly tape both together to make smooth outer covering. (White tape is used here for clarity.) Bottom: As shown here, splice can be made barely 1/16″ larger than the cord itself.

Extension Cords. When an extension cord is necessary for an appliance, match it to the cord on the appliance itself. Caution: never run a cord under a rug. Abrasion will quickly wear away the insulation, creating a fire hazard. All cords should be exposed, and so arranged that access to them is quick and easy at all times.

When adding extension cords, take into consideration the length of the original cord plus the extension, to figure from the chart that follows. Also check connecting plugs, as loose connections can set up high resistance.

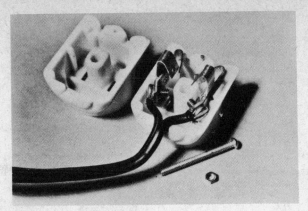

Outlet end-fitting of extension cord is opened by removing screw that holds halves together (above). Cord wires then are attached to terminals.

Above: Lamp sockets that fail to work may have loose connections at terminal screws. Disconnect, remove socket base cap, and tighten terminals. Right: Socket parts consist of (left to right) outer metal shell, inner fiber sleeve, threaded socket, base cap with fiber liner.

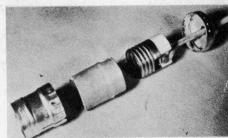

Closet light (right) comes complete with cord and plug. Button switch on base of socket contacts door as it closes, turns off light. When door opens, light goes on.

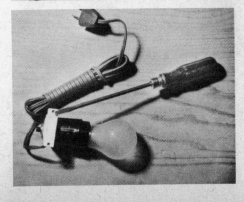

HOW TO SELECT CORDS

Types and usage of extension cords

	Type	Wire Size	Use
Ordinary Lamp Cord	SP SPT	No. 16 or 18	In residences for lamps or small appliances.
Heavy-duty—with thicker covering	S or SJ	No. 10, 12, 14 or 16	In shops, and outdoors for larger motors, lawn mowers, outdoor lighting, etc.

Ability of cord to carry current (2 or 3-wire cord)

Wire Size	Type	Normal Load	Capacity Load
No. 18	S, SJ or SP	5.0 Amp. (600W)	7 Amp. (840W)
No. 16	S, SJ or SP	8.3 Amp. (1000W)	10 Amp. (1200W)
No. 14	S	12.5 Amp. (1500W)	15 Amp. (1800W)
No. 12	S	16.6 Amp. (1900W)	20 Amp. (2400W)

Selecting the length of wire

Light Load (to 7 amps.)	Medium Load (7-10 amps.)	Heavy Load (10-15 Amps.)
To 15 Ft.—Use No. 18	To 15 Ft.—Use No. 16	To 15 Ft.—Use No. 14
To 25 Ft.—Use No. 16	To 25 Ft.—Use No. 14	To 25 Ft.—Use No. 12
To 35 Ft.—Use No. 14		To 45 Ft.—Use No. 10

NOTE: As a safety precaution be sure to use only cords which are listed by Underwriters' Laboratories.

CHAPTER NINE

WALL AND CEILING FIXTURES

You will seldom encounter complications in mounting wall or ceiling fixtures as most of them are designed to fit standard outlet boxes. In general, both wall and ceiling fixtures may be mounted either by means of a threaded "fixture stud" attached to the center of the box, or by a mounting strap with boxes that lack the stud. The stud is favored for heavier fixtures.

Fixtures in New Work. In new work (home construction) the boxes are mounted in the framing before the walls and ceilings are covered. Standard metal hangers are then used to secure the boxes to studding or joists. A wide variety of these hangers is available at hardware and electrical supply dealers. Some are designed to hold the box firmly to a single stud; others are adjustable to permit spacing the box at any desired distance between studs or joists. As the house framing is exposed, installation of the boxes is very simple.

Fixtures in Old Work. In old work a number of different methods and devices are used to simplify installation. Switch and outlet boxes, for example, can be mounted without fastening to the wall framing at all. The plaster or wallboard is used as support. One type of box used in this way is fitted with a folding clamp on each side. It is inserted in the hole cut to fit it and pushed in so that the front brackets seat firmly against the outside of the wall. The side clamps, flat against the box, slide into the hole easily. To lock the box in place, the screws that open the clamps are turned in, spreading the clamps and drawing them snug against the inner surface of the wall.

Another device that can be used on standard boxes is made of thin sheet metal and sold under such trade names as "Hold-It." In form it might be compared to a T with two vertical legs instead of one. The box is pushed into the hole made for it and brought up snug against the outer wall surface (stopped there by the brackets or ears at the ends). Then the sheet-metal fasteners are slipped in on each side of the box, tipped at an angle with the cross of the T inside the wall. Next, they are set straight (so the T cross is vertical inside the wall if the box is set vertically) and pulled outward by the legs until the cross member is seated firmly against the inner surface of the wall. Then the protruding legs are bent over the edges of the box to lock it in place. From there on the box is handled in the usual manner.

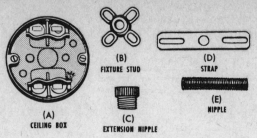

(A) CEILING BOX	(B) FIXTURE STUD	(D) STRAP
	(C) EXTENSION NIPPLE	(E) NIPPLE

Components of ceiling box and ceiling-fixture mounting.

If there is no stud in box, strap is fastened to threaded ears, and fixture is then fastened to strap.

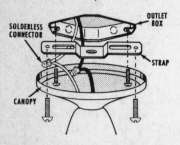

If box has stud, strap may be mounted with threaded nipple and locknut. Fixture is mounted on strap in usual way.

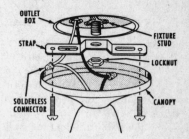

Ceiling drop fixtures are usually mounted from stud, using two nipples joined by "hickey." Both hickey and collar can be adjusted to draw fixture canopy snugly against ceiling.

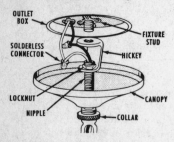

Wall fixture is attached to rectangular box by means of strap and nipple.

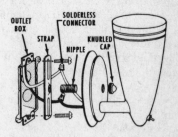

Outlet box with central stud requires only adapter and nipple to attach wall fixture.

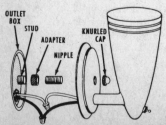

Fluorescent fixture can be ceiling-mounted with stud, nipple, and strap.

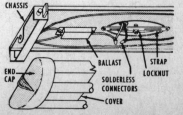

Montgomery Ward & Co.

Ceiling Boxes. To mount a ceiling box an "old-work hanger" is used. This is simply a metal bar or heavy strap with a sliding stud on it. As a first step, cut a hole in the ceiling to match the size and shape of the box to be installed. The cable may then be snaked through the hole and connected to the box. (The box must be a type with a knockout or hole for stud-mounting.) Next, slip the bar or strap of the hanger up through the hole and slide the stud so that half the hanger extends to each side of the hole. The stud should be centered in the hole. Then push the box up into the hole so that the stud extends downward through the stud hole in the center of it, and turn the nut loosely onto the stud. Make sure the cable is not cramped against the inner ceiling surface, then tighten the stud, and the job's done. The bar should run crosswise of laths in plaster ceilings.

If the ceiling box is to be mounted in a ceiling with an unfinished attic above it the job can often be done simply by using a new-work hanger. Cut the hole in the ceiling first, then mount the hanger to hold the box in the hole, working from above in the unfinished attic. A few boards or half a sheet of plywood can provide a working platform across the ceiling joists if there is no flooring in the attic.

Mounting a New Fixture in an Old Box. If you are simply mounting a new fixture on an old box, the work is much simpler. An examination of the existing fixture will quickly tell you whether it is mounted with screws or a stud. If you can arrange to have a helper in removing or mounting fixtures, the job will be easier and faster. If not, have some heavy but soft iron wire on hand, like that used in coat hangers. This can be bent to form a double-ended hook to hold the fixture while you disconnect the wiring.

All such work, of course, must be done with the current definitely off. You can turn it off at the main switch or unscrew the fuse that supplies that particular circuit—if shutting off power throughout the house would involve serious inconvenience. But do not merely shut off the wall switch supplying the fixture. There is too much danger of someone inadvertantly turning it on again while you're working on the wiring.

Form one end of the wire hook before you remove the stud nut or screws holding the fixture in place. If you have a helper, of course, he can hold the fixture while you disconnect it, eliminating the need for the hook. Once the fixture nut or screws have been removed, lower the fixture carefully to provide room to disconnect the wiring. Almost always you will find a hole or clamp in the box over which the hook can be anchored. Do not anchor it over the wires. The lower end can be slipped through the stud hole or the screw holes of the fixture canopy and bent upward for a firm hold. This prevents the fixture from slipping or dangling in such a way as to place a strain on the wires before you can disconnect them.

If the wires are connected with solderless connectors, the job is fast and easy. If the wires are spliced, soldered, and taped, clip them close to the splice on the fixture side of the splice, leaving just enough fixture wire to grip with pliers. With the fixture disconnected and out of the way, you can use a soldering iron to melt the solder on the splice while you pull off the remaining fixture wire with pliers. As the fixture wire is usually the flexible multi-strand type, the solid wire from the box may have little or no distortion in the bared end-portion, making it usable again without clipping. Try to save as much of its length as possible for ease of work when you resplice.

Wiring the Fixture. Merely connect the white wire to the white, black to black, as in regular outlet wiring. In connecting the new fixture use solderless connectors if you want to speed the job and simplify any future changes.

Because of the wide difference in fixture design, you may need a different length nipple for the new one if the mounting is by means of stud and nipple. Often, however, you will find that the "hickey" which joins the nipple to the fixture stud in the box will provide enough adjustment. This is shown in the drawings.

Fixture Repairs. If a fixture fails to light when switched on, the trouble may be either in the fixture or the switch. Often a defective switch can be spotted by the sound it makes. The click (if it's a snap switch) may sound dead. If the switch controls more than one fixture, however, and the others light, the trouble is in the individual fixture. If in doubt, try replacing the switch first. It's an easier job than removing a fixture that may not be in need of repair.

The commonest causes of fixture troubles are in switches mounted on the fixtures themselves, and in the bulb sockets. Naturally, if one or more bulbs fail to light in a multi-bulb unit, check the bulbs first, trying them in another lamp or fixture. Next, *with current off* at the fuse or main switch, check the center contact at the base of the socket, using a flashlight. If the contact is blackened, scrape it clean with a sharp-tipped screwdriver. Then try a new bulb in it. Often corrosion at this point is the cause of the trouble. With current on again, the bulb should light if the contact was at fault.

If the bulb still does not light, the trouble is either a broken wire or a loose connection at the socket terminal screws, or possibly a defect in the socket itself. With current off, you can remove the socket from its base (if standard) by pressing it at the point indicated by the word "press" stamped on the shell of many popular brands. If the terminals are tight, disconnect them and substitute another socket as a test.

With current on, if a bulb lights in this socket you have located the trouble and can then buy a new socket to match the fixture, if the test socket was borrowed from a lamp or from the workshop gadget box. If the test socket bulb does not light, the trouble is in the fixture's wiring. (Neon-glow test lights are sometimes used to check the various parts of fixture wiring, but socket substitution is safer as no live terminals are exposed when the current is on.)

If wiring must be replaced, particularly through curved ornamental tubing (as in many chandeliers) it is best done by firmly twisting the bared ends of the new wiring to those of the old, so the old wiring pulls the new wiring through. As you pull the old wiring out of one end of the tube, you pull the new wiring in the other end, and finally all the way through.

Soldering Electrical Work. If a splice must be soldered or heated with a soldering iron to separate it, you can use an old-style soldering iron heated by a blow torch. Naturally, the current to the wiring being repaired must be off. Certain types of electric soldering irons which have no electrical contact between current supply and heated tip can be used with an extension cord to another circuit in the house that has not been shut off. Before doing any electric soldering by this method, however, *be sure* to check with the manufacturer of the soldering iron as to its safety in this type of use. The reason: when a fuse is unscrewed to shut off current to a house circuit only the hot wire (black) is "broken." The white ground wire is still connected to the power source. Any contact between the white wire of the fuse-out circuit and the black (hot) wire of any other circuit would cause a direct short.

FACTS ABOUT LIGHTING. In planning any fixture replacement try to combine lighting efficiency with decorative effect. Many of the classic fixture and chandelier forms, long abandoned, have returned to popularity in modern form. Many have features previously unavailable, providing various types of lighting at the touch of a switch, ranging from a soft glow to a bright, shadowless illumination.

For general overall lighting in a room, the ceiling fixture as we know it today can provide shadowless light in working areas such as kitchen or workshop. This is achieved by using several fixtures spaced apart to flood a given area with light from a variety of angles, like the shadowless lighting long used in hospital operating rooms. The principle is simple: a shadow cast by light coming from one direction is illuminated by light coming from the opposite direction. The effect is especially easy to achieve with several large fluorescent fixtures. Other fixture types with various additional features such as high-low switches offer a light range from a mere candle glow to full illumination. Others by Lightolier have a three-way switch, enabling you to have focal downlight, candlelight, or a combination of both. Some models have a concealed spotlight to dramatize a centerpiece on a formal dining table. With all these features to choose from, providing good lighting is easier than ever.

Since the ceilings in today's homes are considerably lower than those in older ones, many people think that chandeliers are out of the question, except perhaps over a dining table where there would be no danger of bumping your head. Such is not the case. The new fixtures have been designed with this in mind, and modified versions that still retain traditional design (where it is desired) are available. Some new designs are conceived so that they fit snugly against the ceiling, and proper proportions in the scaled-to-fit models make their use practical in the smallest homes.

Other types are mounted on ceiling tracks. This allows you to have the overhead light at more than one location, according to your needs.

The Science of Lighting. Good lighting does not mean just enough light. It means also control of glare, which can make reading, sewing or other tasks difficult, and irritate the eyes. Proper distribution of the illumination is also important. Unshaded lamps do not produce good lighting by any means. Use lamp bulbs of ample wattage with a proper shade to focus the light without allowing direct light to assail your eyes.

To dramatize a lovely bedroom, proper lighting is essential. Here, a cornice across each window conceals an incandescent lighting strip that provides a dimmer-controlled "wash of light" on the drapes. A pair of crystal sconces, a tall white lamp between the beds, and soft lighting in the wall niche between the windows complete the lighting design.

Light from recessed wall fixtures, controlled by a dimmer switch, is reflected by the white background of the curtains. White table lamp provides local light beside sofa.

Surface-mounted ceiling lights dramatize a room where paintings need spotlighting. The same type of fixture adds charm to a corner reserved for music, enhances the pattern of the drapes.

A chandelier, used with a dimmer control, lights this dining room to perfection. The strip-lighting behind the window cornice, reflected by the curtains, highlights the wall mural.

Formed of thousands of tiny ceramic beads, this ceiling-hugging fixture is convenient in low-ceilinged rooms where overhead light is desirable but where a hanging fixture would be in the way.

Lightolier, Inc.

So shallow that it looks built-in, this ceiling fixture takes four lights, with a maximum of 75 watts each.

When your lighting, by experimenting with various lamp wattages and shades, seems comfortable, and your eyes do not tire after extended use with the lamp, take a reading with a photo light meter, and make a note of that reading. Make the reading on a piece of ordinary gray cardboard. Then when you want to duplicate your successful formula in another part of the house, all you have to do is try different bulbs until the meter reading is the same as before. The reason for using the gray cardboard is that different colors on walls and ceilings have different reflective factors, but the neutral gray assures you of getting the same effect each time. This is an especially good procedure for desk and reading lamps, not too important in over-all illumination.

About Lamp Bulbs. Larger lamps give more light per watt of power used than smaller sizes, and are therefore more efficient. For instance, three 60-watt lamps with a total wattage of 180 give 10 percent more light than five 40-watt lamps, with a total of 200 watts. A 100-watt lamp provides 1670 lumens. Two 100-watt lamps provide 3340 lumens, but you get 5680 lumens from a single 200 watt bulb.

CHAPTER TEN

INSTALLING AND REPLACING SWITCHES

THE TWO wires in regular 120-volt house wiring might be compared to the straightaways of a long oval highway with a steady stream of traffic flowing around it. If you open a drawbridge at any point along it, traffic must stop all the way around. Thus, in house wiring, we need "break" only one wire to stop the flow of current. For the current is literally on a round-trip course— out along one wire, back along the other.

Installing a Simple Switch. The black wire is always the one broken by the switch, never the white wire, as the white one is grounded to the earth and must always have a continuous run back to the grounding point for safety, as explained earlier. In the simplest switching arrangement, a switch is inserted into the line running to the light or appliance. The incoming cable is attached to the switch box as described in the chapter on house wiring. The outgoing cable to the light or appliance is attached in the same way. About 8″ of free wire should extend into the box from both.

To connect the wires, bare all the ends and splice the two white ones together. Connect the two black wires to the two terminal screws of the switch. Simply fasten the switch to the box with the screws provided at the ends, attach the cover plate, and the job is done.

It is often impractical or inconvenient, however, to connect the switch directly into the run of the wiring to the fixtures it is to control. You may, for example, want a switch at a considerable distance from the run of the line. If so, there is a very simple way of doing it.

Instead of connecting the black wires directly to the switch, splice each one to the end of one of the wires in a 2-wire cable, and lead that cable to the switch location. There, the other ends of the cable wires are connected to the switch. As you cannot buy cable with two black wires, however (2-wire cable has one black wire and one white wire), you must connect one of the black wire ends to a white wire in the cable. This is the *only* wiring situation where a black wire is ever connected to a white one.

Wiring arrangement for adding a wall switch to control ceiling light at end of run.

To add wall switch to control ceiling light in middle of run.

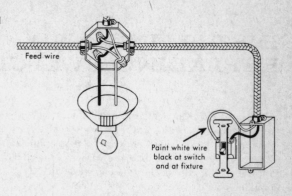

Feed wire

Paint white wire black at switch and at fixture

Feed wire

Paint white wire black at switch and at fixture

To install two ceiling lights on the same line, one controlled by a switch.

To add a switch and convenience outlet in one outlet box beyond existing ceiling light.

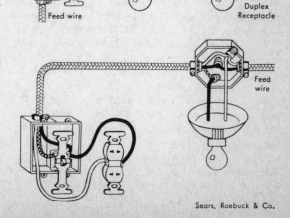

Red wire

Feed wire

Pull Chain Light **OR** Duplex Receptacle

Feed wire

Sears, Roebuck & Co.

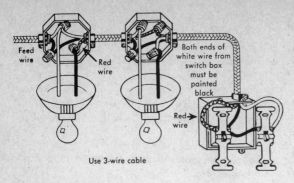

Feed wire

Red wire

Both ends of white wire from switch box must be painted black

Red wire

Use 3-wire cable

To install one new ceiling outlet and two new switch outlets from existing ceiling outlet.

Whenever this type of connection is made, the ends of the white wire *must* be painted black both at the splice location and the switch location. In other switching arrangements the black-to-white connection may occur also at a lighting fixture. Here, too, the white wire that is being used as a black one must be painted black. The paint eliminates the chance of dangerous errors in future wiring repairs or alterations.

3-Way Switches. This type of switch is used to control a light from two different locations. For example, you can turn an upstairs light on from the foot of the stairway, then turn it off from another switch at the head of the stairs. Yet it can be turned on again from the downstairs switch without changing the position of the upper switch that turned it off. In short, it permits you to turn a light on or off from either switch regardless of the position of the other.

How the 3-way switch works. Typically, the 3-way switch has a pair of brass terminals and a single darker one, often copper colored. The dark one is called the "common" terminal. As there is some variation in the identifying systems, be sure to find out which terminal is the common one when you buy this type of switch. You'll need to know when you connect it.

The common terminal is attached inside to a switch arm that can connect it internally to either of the other two terminals, depending on which way the outside switch lever is pushed. The diagram of the principle on which the 3-way switch circuit works shows the wires widely separated for clarity. Actually, all three wires run inside a single 3-wire cable. The individual wires

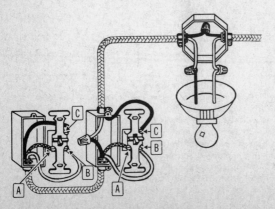

Three-way switches controlling outlet located beyond both switches. A is red wire, B white wire, C black wire.

are colored red, white, and black. In all work using this type of cable, the white wire must be painted black at both the switches and the fixture if it is used as a black wire, that is, as a switching wire.

In the diagrams of the many switching arrangements possible with the 3-way switch, the black wire is shown black, the white wire, white, and the red wire cross-hatched. In the actual installation, the white wire would be painted black where used as a switching wire.

4-Way Switches. Less common than the 3-way, this switch permits a light or appliance to be controlled from more than two locations. If you want to control the light from additional locations simply use additional 4-way switches.

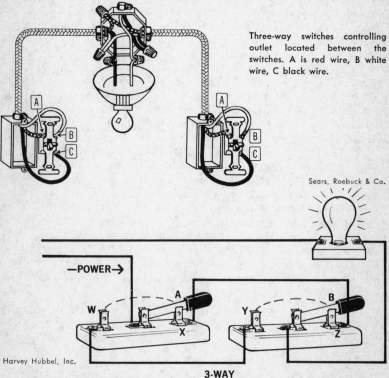

Three-way switches controlling outlet located between the switches. A is red wire, B white wire, C black wire.

Sears, Roebuck & Co.

→POWER→

Harvey Hubbel, Inc.

3-WAY

Simple knife switches illustrate principle of 3-way switch. Power is fed to center (common terminal) of first knife switch. If handle were at W, current would flow to Y but no farther, as second switch handle is at Z. Light would be out. But as the first knife switch handle is a A, current flows through X to Z, then through handle B to common terminal, and on to the light. Light is on. Now note that if you swing either handle to opposite position, the circuit will be broken and the light will be out. Also, if you swing either handle once again to the opposite position, the light will again go on. The same connection changes take place inside a standard 3-way switch.

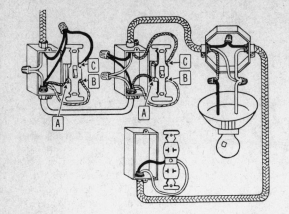

Three-way switches controlling ceiling outlet located beyond both switches. Receptacle is always *hot*.

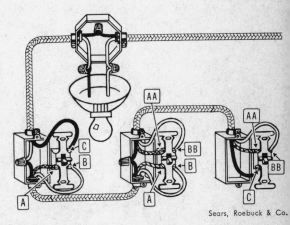

Combination of 3- and 4-way switches is used wherever you wish to control one or more lights from three separate locations. Connect 4-way switch as shown for center switch. Three-way switches are on either side. A is red wire, B white wire, C black wire.

Sears, Roebuck & Co.

Four-wire cable is used in the wiring of 4-way switches. As it is not available at all electrical supply dealers, two separate lengths of 2-wire cable may be used instead. In either case, the ends of any white wire used as a switching wire should be painted black. The only time the white wire is left white is in its commonest use, as the wire carrying the grounded side of the circuit.

Replacing Damaged Switches. When a switch fails do not attempt to repair it. In many cases, after long service, a small spring-metal part breaks and in most cases cannot be replaced. In other instances, pitting of the contacts may cause failure. In either case a makeshift repair can be dangerous, so replacement is the answer.

The first step is always to shut off the current. Next, remove the switch cover-plate by removing the screws that hold it to the box. Then pull the switch, itself, straight out from the box. The replacement switch should be on hand. In most cases you can tell in advance the type it should be. If you are not sure, in a multi-switch arrangement, whether the broken switch is a 3-way or 4-way type, shut off the current and examine it before you buy the replacement.

If you have any doubts as to your ability to connect the new switch correctly, remove the connecting wires from the old switch one at a time, and connect them one at a time to the corresponding terminals of the new switch.

In buying a replacement switch, be sure you select one that is rated at least as high as the old one. You'll find the rating stamped on the metal strap of the switch. Ordinary switches are rated "10 A 125 V—5 A 250 V." This means the switch can be used to control up to 10 amperes if the voltage is not higher than 125 volts. It can be used to control only 5 amperes if the voltage is above 125 but not higher than 250 volts. Switches are also available for higher loads. Check the old one. If you are not sure of its rating, take it with you when you buy—but leave the current off until replacement is made.

If you install a switch in a new location for added convenience, as in replacing a pull chain from a light fixture, remember not to install any switch where it can be reached by a person standing in a bath tub. This applies to both pull chain and other types.

Special-Purpose Switches. If your garage is separated from the house and there is no light switch in the house to control the garage light, considerable work may be required to add such a switch. But it's also inconvenient to walk from the garage to the house in the dark after turning off the light.

There's an easy answer in the *time-delay switch*. This fits in the same switch

Cord switch is inserted in run of lamp cord to provide control from any convenient point. One wire of cord runs through switch, other is cut and attached to two terminal screws. Switch halves bolt together.

box as the conventional toggle switch, and it is connected in the same simple way, using just two terminals. But when you turn off the switch it allows you about half a minute to reach the house before the light goes out.

The *time-clock* switch can be set to turn on lights or appliances and turn them off at any preset times. If you are away and you want your lights to go on at a certain time in the evening it can do the job. But if the power should fail for a period of time and then be restored many of these devices will be out of timing. (A few are spring-clock-operated, winding the clock electrically.)

To assure that the lights go on when darkness approaches, the *photocell switch* is used. As soon as the light level drops to a preset degree, this device operates a relay switch that turns on the lights connected to it. When the light level rises the following day, the lights are automatically switched off. This type of unit is designed so that its sensitive cell can be adjusted to pick up exterior light only; the lights it operates do not affect it. To prevent the unit from responding to sudden flashes of outside light, such as auto headlights or flashes of lightning, a time-delay unit is often included.

Time clock and photocell switches are not connected like ordinary switches. Both incoming cable wires are connected to them, as are both wires of the outgoing cable (to lights or appliances). This is necessary in order to supply the power needed to operate the mechanism in the switch unit. When the switch operates, however, only the black wire is broken by the switch. There is considerable variation in the design of these switches but the terminal screws are either brass and chrome coded or plainly labeled to assure correct connection of black and white wires.

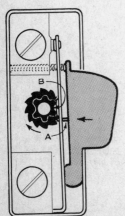 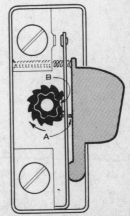

How a press switch works: In "on" position (left), armature connecting terminals is held in contact with top terminal by its own spring tension. Small knob on armature (B) rests between cogs of gear on ratchet wheel. When switch button is pushed, pin (A) turns ratchet wheel, cog presses knob and armature outward, breaking contact at top terminal (right). Switch is then "off."

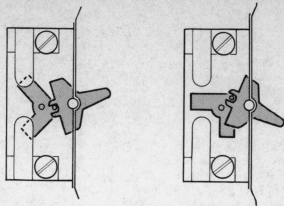

How a toggle switch works: In "on" position (left), L-shaped armature is in contact with both terminals. Armature pin is engaged by twin prongs on switch lever. When lever is pushed down (right), it pivots armature so that contact is broken in both terminals. Switch is now "off."

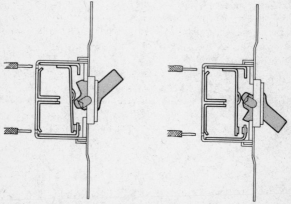

How a silent toggle switch works: In "on" position (left), spring-steel armature rests against bottom terminal, thus establishing contact in circuit. When switch lever is pushed down (right), it presses against small knob on armature, breaking contact in bottom terminal. Switch is now "off."

How a mercury switch works: Switch lever is integral with small, hollow cylinder containing mercury. In "off" position (left and center), contact (small circle) is above mercury, but when lever is pushed up, cylinder pivots and contact is immersed in mercury. This establishes connection between both terminals.

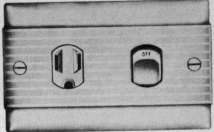

Pass & Seymour, Inc.

Switch plates are available in many combinations: Left, two switches and night light; right, switch and grounded outlet.

Relay unit (left), about the size of a small radio tube, is heart of low-voltage remote-control switching. This lets you run low-voltage (24 volt) wiring from switch to light fixture, operate it from any point desired without need for heavy wiring. Low-voltage remote-control switch is shown above.

General Electric Co.

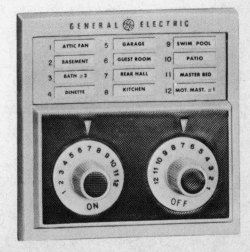

Master selector switch for low-voltage remote control allows you to turn on all lights throughout house or light all rooms along path you intend to take through house.

Parts of Luxtrol electronic light control that permits adjusting lights to any level of brightness desired. It handles up to 500 watts, fits most standard wall boxes.

For simple two-level high and low lighting, you can use GE Hi-Lo switch (left) mounted in ordinary box in place of old snap switch.

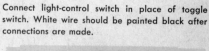

Connect light-control switch in place of toggle switch. White wire should be painted black after connections are made.

To install Luxtrol light control, shut off current and remove ordinary toggle switch from box.

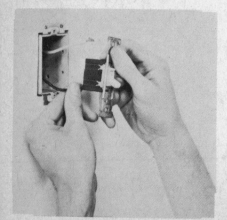

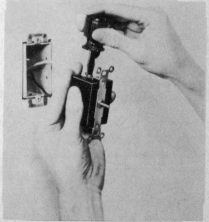

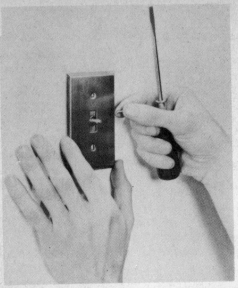

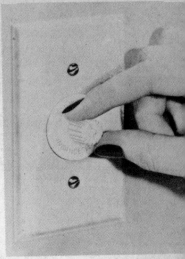

Control dial is pressed onto splined shaft; cover plate painted to match wall—and you can control lighting at turn of dial.

After light control is mounted in box, install cover plate over projecting shaft.

Mercury switches. Designed to operate without the usual click, these switches are useful where quiet is important, as in bedrooms and nurseries, or in nearby locations. They are available in both the standard single-pole type and the 3-way. (The single-pole switch is the basic type with two terminals.)

In using mercury switches it is important that they be mounted according to the manufacturer's instructions and that they be level. As they operate by causing a small amount of mercury to flow from one position to another (opening or closing contacts) they cannot function properly if mounted in a tipped position. Their wiring, however, is the same as conventional types.

Low-voltage switches offer a simple solution to an otherwise extensive wiring problem. This type of switch, in its common form, contains a transformer and relay switch. Simple low-voltage wiring like doorbell wiring can be led from it to switch locations elsewhere. The operating unit itself is placed at the location of the light. Thus, if a particular basement light cannot be controlled from upstairs, you can add the desired control by simply running lightweight bell wire from light to switch location instead of regular wiring cable.

Dimmer switches. Luxtrol, a product of The Superior Electric Company of Bristol, Connecticut, allows you to control light intensity and brilliance. This light level control, in the small size, fits into an ordinary switch box, and is connected like an ordinary switch. In large sizes for higher wattage, a larger box is required. (Similar switches are made by other manufacturers.)

This unit acts like a transformer. Both incoming wires are connected to terminal screws and both outgoing wires are connected to other terminal screws. A dial control replaces the usual switch lever. Bright light for reading, soft glow similar to candlelight for dining, are at your fingertips. Dim to comfortable level for watching TV. Turned very low, an ordinary lamp can serve as a night light for a stairway or bath. This use is also excellent in sick-room or nursery. A built-in switch turns the lights all the way off when the control is turned past the lowest setting.

Before installing special types of switches, however, it is wise to check with your local code. Your electrical supply dealer will be able to advise you, in most instances, if there are restrictions involved in the use of these units.

- security.
- decoration
- safety - Ground fault receptable

CHAPTER ELEVEN

OUTDOOR WIRING

OUTDOOR LIGHTING has been used for a long time for safety, along paths and steps, especially those that are prone to be slippery in ice and snow, or where toys and other obstacles could cause a serious fall in the darkness. But wiring the outdoor living areas for more fun and convenience all year round is something else again, and opens up new vistas for the home electrician.

Outlets designed for the purpose allow the use outdoors of portable radios, television, and hi-fi equipment for entertainment. Electric grills and roasters, toasters and coffee-makers are as easy to use outdoors as inside the home. Exotic lighting effects on lawn and garden areas are an added attraction that will enhance your grounds.

Underground Wiring. Weatherproof outlets and switches, and wire made for underground runs, make outdoor lighting increasingly more feasible. When planning to use underground wiring, be sure to check with your local code as to the type of wire you should use. It is the usual practice to use UF type of wire (underground fused) unless the local code forbids its use. The term "underground fused" does not mean that there is a fuse underground but that there must be a fuse inside the house at the start of underground wiring. Consult your electrical supplier for the type of box to use inside and weatherproof outlet for the outside.

The most common use for underground wiring is to lead it to a lamp-post or lantern to light the driveway. The underground cable should be buried at least 18″ deep to prevent damage from frost or deep-cutting garden tools. If lawn area has to be disturbed when the trench is dug, use a spade to lift the turf, and lay each square carefully right-side up along the trench so that roots are not exposed to drying sunlight. Use a long-handled shovel to remove the earth to the desired depth and pile it on the opposite side of the trench. You are now ready to lay the wiring.

After the wire or conduit is installed, fill the trench with earth and carefully replace the squares of turf. Gentle tramping by foot, followed by watering, will eradicate all signs of damage to the lawn.

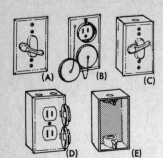

Time switch, permanently mounted, can turn on driveway lights in advance when you plan to return home late.

Weatherproof switches and outlets for outdoor wiring: A. Flush-mounted weatherproof switch. B. Flush-mounted weatherproof outlet. C. Surface-mounted weatherproof switch. D. Surface-mounted weatherproof outlet. E. Metal weatherproof box.

Portable time switch can be used with plug-in equipment to provide light for driveway or path at proper time for late-returning residents.

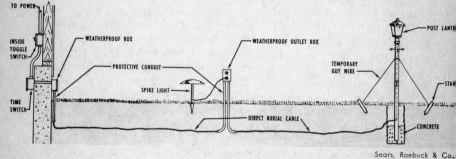

Sears, Roebuck & Co.

This drawing shows how a typical underground wiring problem can be handled. If you use UF type of wiring, a fuse must be provided inside the house at the start of the run.

Weatherproof Outlet Boxes. These can be mounted on posts or conduit stubbed up through the ground to allow for spike lights and other conveniences along the run from house to lantern post. Consult your local code for this. Weatherproof outlets spaced along the house-wall of porch or terrace can be used for portable tools, grills, etc. Under eaves and overhangs, when coupled with bulbs made for the purpose, they provide illumination and tend to keep prowlers away.

Dramatic Effects. Do not limit the lighting on your patio or terrace, use it for dramatic effects. Aim one light to shine through the branches of a favorite tree for a romantic atmosphere. A low-placed light, with its beam skimming the surface of the lawn, picking out a flowering shrub or garden, will surprise you. It can completely transform the appearance of the plot as it accents and dramatizes the plantings so that they look entirely different from the way they appear under overhead sunlight. Play around a little with the direction of the beam until you achieve just the desired design of light and shadow.

If you have a large garden, with walks, paths, and steps for different levels, spike lights will enhance your garden at night. Colors may be intensified or lightened by controlled lighting, and the various shades of green found in all garden areas will stand out as in a stage setting.

Ready-Made Fountains. With outdoor wiring even garden fountains are not a luxury today. Garden shops and department stores have them at budget prices, and your own imagination can put them to work to suit your garden decor. The cool sound of tinkling, splashing water is refreshing, and when there is play of light over the spray at night (try colored lights for a fantastically beautiful effect), you will have a display of distinction and charm.

Outdoor Fans. When an outdoor party is planned for the terrace or patio, and the weather turns hot, humid, and stuffy with no breeze to alleviate the situation, be smart, and use a fan. The whole party will be more comfortable, as any movement in the air will relieve the stuffy, oppressive atmosphere and, as an added bonus, drive insects away. Have a yellow "bug-light" near you, and a bright white light at a considerable distance, with a pan of kerosene or insecticide below. The "bug-light" will not attract insects, but they will be drawn to the brilliance of the white light. This will work any time, but when humidity makes the bug problem worse, as it usually does, the fan does double duty. It pays to take every advantage of your outdoor wiring, and the longer you use it, the more ideas come to you.

Lighting for Fall and Winter. Don't consider this outdoor lighting for the summer months only. The brilliance of fall foliage can be enjoyed in the evening by people whose work takes them to the city by day, and as this is a fleeting season it pays to take advantage of your lighting to enjoy it. Also it is fun to find different lighting arrangements to suit the changing colors of the leaves of fall.

In winter, enjoy a snow scene to its fullest. Picking up the sparkle of falling snow can be rewarding in more ways than one. Not only is it a pleasure to watch from the cozy warmth of the fireside, but as you watch you can estimate the accumulation of snow. This will alert you as to when to break out the snow-blower, so that you don't have a tough job in the morning when you have to get to work or an appointment. And if using the snow-blower before retiring is indicated, as when the fall is heavy and swift, or bad drifts seem imminent, the outdoor lights will ease your task, as not too many snow-blowers are equipped with headlights. Thus, with the major job done, a light going over in the morning will be all that is necessary, and you can relax for a good night's sleep without the thought of rising in a cold bleak dawn with a big snow-clearing ahead of you before starting out for your job.

Outdoor wiring is also fun if you have a skating rink on the lawn in winter. Not only can you light the skating area, but you can use all your usual patio equipment to great advantage. Set up a simple table for a buffet, and with your electric percolator, toaster, and grill, hot drinks and food are ready for a crowd of hungry skaters at all times. An electric fry-pan or casserole can keep a main dish ready to serve, and for the younger set, the grill can provide franks or hamburgers with no fuss and they will enjoy it as a lark—and they won't be tracking snow into the house. A portable electric heater can take the chill out of fingers and toes, and the ruddy glow from a bowl-type heater simulates the glow of a fireplace for added atmosphere.

For lighting game areas at night, 150-watt flood and spot bulbs mounted in trees and buildings can produce the proper light. Experiment with angles to provide light without glare.

For a safe entry, place outdoor lights so they do not create glare. They should always have an adequate shade, as the blinding effect of a bare bulb is almost as dangerous as no light at all.

Careful location of lighting fixtures adds charm as well as safety to garden walks and steps, and dramatizes the entry to the garden or patio area.

A few well-spaced outdoor lights can keep the area outside your picture window aglow at night for added enjoyment.

Westinghouse Electric Corp.

A subtly illuminated patio adds charm to a summer night, but do not over-brighten the area so that guests feel as though they are in the spotlight.

Artful lighting blends the interior and exterior decor of this house. Careful attention to prevent glare on glass or screen does the trick.

Westinghouse Electric Corp.

A garden successfully illuminated at night produces a charming effect. Trees are highlighted by strategically placed lights, with the bulbs well shielded.

The home skating rink may sound like a luxury item for the homeowner with a tennis court to flood. Such is not the case. It is a simple do-it-yourself project for the fellow who has outdoor wiring at his disposal. Weatherproof outlets, a lawn area level enough, and a properly protected outside faucet are all you need. Proper protection for a faucet to be used in such a case means one protected with electric heating tape, preferably tape with thermostatic control.

Electric Heating Tape. In cold parts of the country especially, electric heating tape can save the homeowner many a costly headache. For instance, if you want to make your outdoor skating rink, a simple form of 2-by-4s can be filled from an outside faucet *if* that faucet is wrapped with the heating tape. Weather cold enough to freeze the water in the form is also cold enough to freeze the water in the faucet, causing the faucet, and the near part of the water line leading to it, to freeze and burst. The damage can be considerable. Don't think that because the faucet is going to be used for only a few minutes that it will not freeze up. It can reach the breaking point in a very few minutes, so be sure to use the tape.

As soon as you are through, disconnect the hose and take it indoors, say into the cellar or garage so it too will not suffer from the cold temperature. If you intend to use the faucet at a later date in cold weather, leave the tape connected, with the thermostat set for protection. If you intend one-time use only in cold weather, close off the water to the faucet from inside the house, thoroughly drain the faucet, and only then disconnect the tape. The tape should, of course, be connected into one of your weatherproof outdoor outlets.

One of the most money-saving uses of electric heating tape is the prevention of ice-dams on the roof. Ice-dams cause damage not only to the roof, but to ceilings inside as well. Frozen downspouts and gutters, which extend beyond the portion of the roof that gets heat from the house, prevent runoff, causing melting snow to back up under shingles and roofing and drain into the house. The damage to interior walls and ceilings can mean a costly repair bill and redecorating as well. Prevention is cheap and easy with the use of the heating tape.

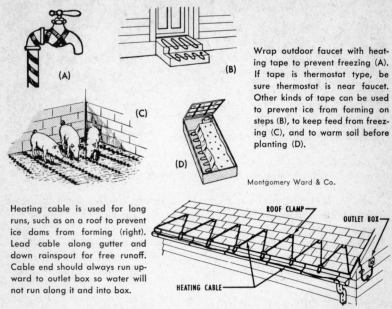

Wrap outdoor faucet with heating tape to prevent freezing (A). If tape is thermostat type, be sure thermostat is near faucet. Other kinds of tape can be used to prevent ice from forming on steps (B), to keep feed from freezing (C), and to warm soil before planting (D).

Montgomery Ward & Co.

Heating cable is used for long runs, such as on a roof to prevent ice dams from forming (right). Lead cable along gutter and down rainspout for free runoff. Cable end should always run upward to outlet box so water will not run along it and into box.

Installed on the roof, in the gutters and downspouts, it will not discolor or mar the roof. Start the cable at a weatherproof outlet, and zig-zag it at a 60-degree angle to the roof edge. Return in the gutter, and drop the end to the bottom of the downspout. Allow the cable to loop *below* the edge of the roof to provide a heated drip point. Manufacturers recommend that a 6-foot cold lead be run *upward* to the outlet so water will run off before reaching the plug, and that you ground metal gutters and spouts with a driven ground. Follow the instructions that come with the tape you buy, and consult your local code before installation. The principle is always the same when it comes to the connection of the tape to the weatherproof outlet: provide a deep sag below the outlet so that water will drip off below the outlet instead of into it, as that might cause a short.

When electric heating tape is used in concrete steps to prevent ice formation (a great safeguard), the cable should be buried according to the manufacturer's instructions. Be sure to get the correct type for this purpose, as this is not the kind used on roofs. Another place for the buried type is across the sill of the garage, where frequently an overhead door can freeze shut in severe weather. If you want a snow- and ice-free driveway, the proper cable can melt the white stuff away. Be sure to buy the correct type and install it properly.

Some types are used by home gardeners in hotbeds to give their plants a head start. It pays the home electrician to investigate all the possibilities of heating tape, as it can solve problems and also suggest new uses to the hobbyist and home-workshop craftsman. If you want specific details about what heating tape can do for you, write to one of the leading manufacturers in the field, such as The Smith-Gates Corp., Farmington, Connecticut.

CHAPTER TWELVE

TESTING WIRING

When you complete a wiring job, especially a fairly extensive one, you can spot any troubles while they are still easy to correct if you make a few low-voltage tests. In fact, the tests can be made when only the wiring (not the switches and outlets) has been installed.

A pair of dry-cell batteries and a flashlight bulb are all you need. If the old-style bell batteries are not available, you can use *one* of the popular lantern batteries. Whatever the type, there must be terminals to which wires can be connected and the bulb must match the battery voltage.

The wiring, of course, must be dead insofar as the regular house current is concerned. Connect the battery or batteries to the black and white wires at the beginning of the wiring run. Splices in the wires along the various runs need not be soldered at this point. And if they are soldered it is better if they are not yet taped, as this will make it easier to track down trouble if any turns up. At outlet boxes (without outlet receptacles) incoming and outgoing wires should be hooked together lightly, white to white, black to black. Don't twist them together or you will distort the bared ends too much to form the loops that will later be attached to the terminal screws. You merely want to make temporary connections where permanent ones will be made later.

To make the tests, touch the threaded (or smooth) outer metal shell of the bulb base to one wire and the center contact at the bulb bottom to the other. At most auto-supply dealers you can buy a socket to fit many small bulb types. This makes the testing simpler, as you can attach short wires to the socket and simply touch these to the wires in the outlet boxes to make the tests.

If the bulb lights you know you have sound connections and unbroken wiring to that point. At switch boxes also connect incoming and outgoing wires, as in the outlet boxes, so current will flow through as if the switch was in the "on" position. If the bulb fails to light at any test point, check over your temporary connections. If these seem sound, reconnect the battery at the other end of the final run of cable involved. Although it is not at all common, a broken wire is sometimes found even in a new piece of cable.

After all outlets and switches have been installed and all splices soldered and taped (or joined with solderless connectors), repeat the bulb test at the outlets before connecting the power. The easy way: connect the little bulb socket to a short piece of lamp cord and a plug.

One of the most important bits of information the testing will provide occurs at the outset when you connect the battery to the start of the wiring. If there is any spark, *do not* make the connection. Instead, connect *one* wire from the battery to *one* wire of the cable. Connect the other wire from the battery to *one* wire from the bulb socket. And connect the other wire from the socket to the remaining wire of the cable. If the bulb lights, there is a short somewhere in the wiring. To locate it, first make certain that no bared wire ends are touching the metal boxes, and that no black wire is in contact with a white wire. If you find any of these conditions, correct them. If the short still persists, disconnect the incoming and outgoing wires from each other in the boxes and test each individual run of wiring by the same method that detected the short. This will show which run is causing the trouble. Although shorts are uncommon in runs of new cable, they can sometimes be caused by careless handling, as when insulation is unknowingly scraped from a wire at a point where it can contact another or a metal part tied to the other side of the circuit. Occasionally, too, it may result from defective cable.

Testing the Grounding. Another test should be made on the grounding of the system, as on armored-cable systems or nonmetallic-cable systems carrying a bare metal ground wire. In addition to making the bulb test between black and white wires at each outlet and switch box, make one between the black wire and the box itself. As both the white wire and the grounding wire or cable armor are connected to the ground at the service entrance, the bulb should light. On a long run of wiring it may be somewhat dimmer, as the armor is not as good a conductor as the wire itself.

The advantage of low-voltage testing lies in the fact that it can be done while the wiring is still open at boxes and easy to repair. And, if a short or other trouble is present, no wiring damage or blown fuses will result. Keep the testing bulb and socket after the tests. You can use it for many other electrical sleuthing jobs, such as testing switches and appliance cords. It permits you to test wiring items before they are put into use. To test outlets and other wiring points with current on you must use a neon-glow lamp tester with insulated prods. This type is also used to detect blown fuses when the appearance of the fuse does not tell the story.

- regular transform
- + p. -rated transformer - protected from overheating
- watt-miser bulbs +
- energy saving ballast,
- low temperature ballast
- dimming ballast

CHAPTER THIRTEEN

FLUORESCENT LIGHTS

FLUORESCENT LIGHTING fixtures are no more difficult to install than ordinary ones, though their larger size requires a certain amount of space planning. They are commonly available in lengths from 18″ to 60″, with a wattage range from 15 to 60. The 48″ 40 watt is one of the more popular sizes. They are also available in circular form. The sizes are based on the length of the tube, which is matched in size by a metal box or channel containing the electrical units necessary to light the tube.

The fluorescent light provides far more illumination per watt of electricity than an ordinary bulb, and it lasts much longer. The life of the fluorescent lamp depends largely on the number of times it is turned on and off. If left on and burning continuously, it will outlast a bulb 8 to 15 times. If turned on and off with the usual frequency in household use, it will still outlast a bulb 3 to 5 times. All fluorescent lamps are not suited to low temperatures, however, so if you plan to use one in a chilly spot like an unheated garage, check with your electrical-supply dealer for the right type.

How Fluorescent Lighting Works. The fluorescent lamp produces its light by means of an arc that jumps from one end of the tube to the other, producing ultraviolet light. If the tube were of clear glass, you would see only a dim violet glow. But when the ultraviolet light strikes the tube's inner coating (often called a phosphor), it causes it to fluoresce, or glow brilliantly. Though the tube has small filaments at each end, these are lighted only for a second or so when the lamp is first turned on. From then on, the arc passing through gas and mercury vapor in the tube is the light producer.

Inside the metal channel box that supports the tube there is a "starter" and a "ballast." The starter is an automatic switch that opens itself after current has flowed through it for a moment. Then it remains open. The ballast is a winding of wire around a steel core and is designed to do two things: When it is disconnected as power is flowing through it (as when the starter opens), it momentarily delivers a higher voltage than the current originally flowing through it. It also limits the total power that can flow through it.

118

When you turn on a fluorescent lamp, current momentarily flows through the ballast coil, then through the filament at one end of the tube, then through the starter, and finally the filament at the other end of the tube. When the starter opens an instant later the ballast delivers a momentary high-voltage kick powerful enough to send an arc through the gas and mercury vapor in the tube. Once established, the arc continues at normal voltage. As the starter remains open, the current flows only through the arc in the tube from then on.

Aside from the tube, it is seldom necessary to replace any part except the starter—after long use. This is an inexpensive item often made with a plug-in base for easy replacement. In many types of units the top of the starter protrudes through the channel box so it can be replaced without opening the box.

Installing Fluorescent Lamps. The complete units are made in both decorative and utility forms. The latter consists of a simple metal box with knockouts for cable connection at a number of points, and with simple mountings for the tube. Though all parts, including the box, can be bought separately, your best bet to make the job easy is a completely assembled unit. If any part fails at a later date, standard replacements are available.

If the unit is to be used in a basement workshop or other workroom, it can be easily mounted on a ceiling joist or on a plywood ceiling-panel and fed by BX or other cable. A pull-chain switch may be mounted in the unit itself, or a wall switch may be inserted in the feed line. Knockouts are provided for standard small switch types in the unit itself, also for mounting screws to fasten it to wall or ceiling.

The units may also be mounted directly on suitable outlet boxes like other fixtures. The important point: buy a fluorescent unit matched to the type of mounting you plan. The rest of the job is practically the same as for other types of fixtures. It is important, however, that fluorescent units be located where the tube is not likely to be broken by accidental impact, as some tubes contain phosphors that are considered dangerous to inhale.

For use above kitchen counters and in other work areas where multi-outlet-raceway systems are commonly used, a special snap-in unit is available to replace a section of raceway cover. One of the best known is that made by Wiremold. The unit is available with or without a self-contained switch.

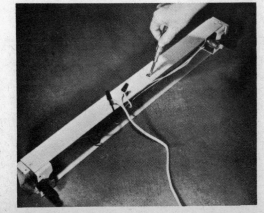

Low-cost fluorescent channel strips are available with knockout that permits use of lamp cord. Punch out knockout with a screwdriver after tapping it with hammer.

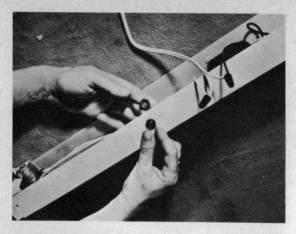

Insulated grommet fits into hole and is locked by nut inside channel box. This protects cord from chafing against metal.

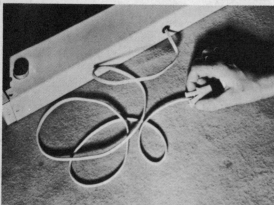

Self-connecting plug is easy to attach. Simply insert lamp cord and press jaws or lever in place.

To replace starter in channel strip, simply remove tube, pull out starter and replace with new one.

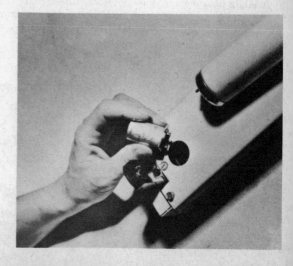

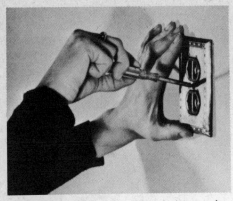

Fluorescent fixture may replace high-mounted outlet. First step: remove outlet cover plate.

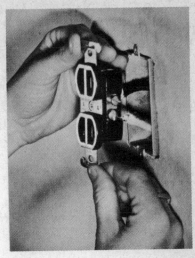

Pull receptacle out of box after removing screws at ends. Then disconnect wire from each side. All this must be done with current off.

Fixture is connected to wires from box using solderless connectors. Fixture wires are shown here coming through fixture's back plate.

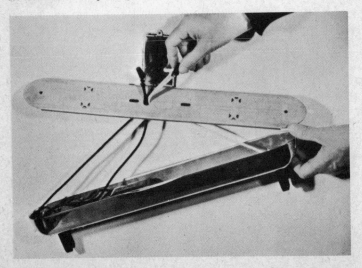

After connections are made, they are eased into outlet box (left). Then fixture's back plate is screwed to threaded ears of box and fixture is attached to back plate. Side flanges of fixture cover exposed areas of box.

Cover of wall unit screw-fastens in place after connections are made (right).

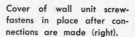

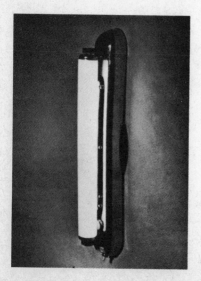

Fixture may be attached to wall (left) in vertical or horizontal position, depending on location. On open wall, vertical position is often preferred, as shown; under cabinet, as in kitchen, it may be horizontal.

CHAPTER FOURTEEN

DOORBELLS AND CHIMES

IN THE past, the doorbell was usually operated by two or four dry-cell batteries providing a total of either 3 or 6 volts. And many of these oldies are still operated that way. Today's doorbells and buzzers commonly operate on 10 volts, and chimes on 16 volts. Whatever the voltage required, you can buy a transformer to supply it, as it will be stocked by the same dealer who handles the bell or chime. If you replace an old transformer-operated doorbell or buzzer with a chime unit, you will undoubtedly have to replace the transformer also, as the old doorbell very likely operated at about 6 volts.

The Transformer. Often no bigger than your fist, the transformer is a small unit. For higher-voltage chime units, however, it is usually larger. In its usual form, two permanently attached wires extend from it, one black and one white. These are the leads of the "primary" or incoming coil of the transformer, and are permanently connected to the 120-volt system of the house when installed.

Some, but not all, transformers are made so they can be mounted directly on an outlet box or junction box with the primary wires running directly into the box. In any event, the wires should run through a suitable connector so that splices are inside the box. The "secondary" wires are the low-voltage ones. These are not attached to the transformer when you buy it, of course, as they must run for considerable distance through the house to link bells or chimes with the door buttons.

Two terminal screws or nuts are provided on the transformer to start this secondary wiring. As the voltage in this wiring ranges from as little as 6 volts to a maximum of only about 20 volts (depending on the chime requirements), it does not have the shock danger of regular house wiring, and does not require the use of cable. Number 18 bell wire, somewhat similar in appearance to lamp cord, is most commonly used. It is generally run along baseboards or other exposed surfaces if the installation is made after the house is built. Insulated staples hold it in place. In new work it may run through the walls.

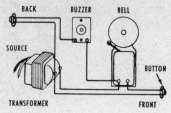

Wiring for doorbell and buzzer. Buzzer signals back door, bell signals front.

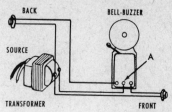

Combination bell and buzzer unit provides separate signal for front and back doors with simplified wiring. Unit has three terminals.

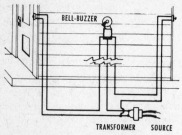

Path of wiring from doors to basement transformer and bell-buzzer unit.

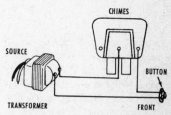

Wiring for one- and two-note chimes, front door only.

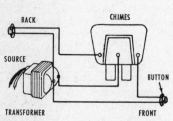

Wiring for one- and two-note chimes, one note for back door, two for front.

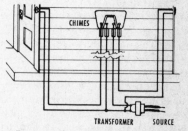

Four-note chime units can sound single note for back door, as many as eight notes for front.

Montgomery Ward & Co.

Wiring a Doorbell. To trace the path of the current, consider the transformer as the source, the doorbell button as the switch, and the chime or bell as the outlet. To lead your wiring to provide a single front-door button or various front- and rear-door arrangements, as shown by the diagrams, you have a choice of bell wire in single-, double-, and triple-wire form.

Although the diagrams show the wires spaced apart to make the circuits easier to follow, you will usually be able to use one of the double- or triple-wire forms. The wiring is run "as is" unless a local code prohibits this. Care should be exercised, of course, to protect the wiring from damage in locations where it might occur. If any problems exist in this regard pick a transformer with overload protection on the secondary (low-voltage) wiring. This is built in to many quality transformers now. With this type the current is cut off in the secondary when trouble occurs.

Note that a separate buzzer for the back door and a bell for the front can be used. On more modern types, a single chime unit can be used with different signals for front and back doors—two or more chimes for the front, one for the back. The difference depends on the make and model. You can also buy a combination bell and buzzer. Note that the single units have three terminal screws. Connect them as shown in the diagrams.

When you install a bell or chime, be sure to locate it where its sound is most likely to reach you anywhere in the house. Remember that the wiring is simple when compared to regular house wiring and that it can be led safely over its course without the work involved in 120-volt work. Keep in mind, too, that even though a particular chime may have provision for an elaborate hook-up, you can usually use it in a simple single-button front-door system.

Testing for Doorbell Failure. To find the cause of bell or chime failure, simply use your low-voltage circuit tester (described under Testing Wiring). In most cases the failure occurs at the doorbell button because of mechanical (usually spring-metal) failure or because corrosion from weather has affected the contact areas. In some cases, the bell will fail to ring at all. In others it may continue ringing after the button is released. In the latter case, remove one of the wires from the secondary terminal screws.

One of the first steps in checking bell or chime failure should always be a check of all visible sections of the wiring for mechanical damage. After doing this, use the tester to check each section of the wiring. In general, you are more likely to find a break than a short.

In certain parts of the wiring a short acts as a switch (or bell button) and causes ringing of the bell. In other areas it simply shorts the transformer and causes the protecting device in it to cut out. So let your first test be this: disconnect one wire from the transformer's secondary screw terminals and touch your tester to the terminal and the disconnected wire. If there's a short, the bulb will light. Be sure the bulb you use matches the voltage of the transformer. Bulbs of lower rating will burn out, those of higher rating will be dim.

The easiest way to test the doorbell button is by connecting it from one transformer terminal through the bulb tester and back to the other terminal. If it works when you push the button, the bulb will light. If it doesn't work, you can clean the contact points in some types with fine sandpaper. This may be the cause of the trouble. If the contacts are not readily accessible, do not try to pry the button apart. Simply replace it.

If the button proves sound and the trouble appears to be at the bell or chime, check the terminal connections. Vibration sometimes loosens them. If the problem is in the chime or bell unit the chance of repairing it depends largely on the individual make and model. A burned out solenoid (not common) may or may not be replaceable in a chime. A broken wire or connection, if it can be detected, can often be repaired. In general, however, failure of the chime unit itself is not likely. In the case of bell or buzzer failure, the low cost of the unit makes replacement the best answer.

If you are planning to install a completely new bell or chime system, buy all of your materials at the same source to assure voltage matching. (You will need to measure for the length of wire needed.) In many cases the entire system is sold as a kit with enough wire to take care of the average installation. If you buy a kit it pays to measure your wiring run in advance in case the length of wire provided is not enough. It is much handier to buy the length of wire required at the outset than to have to buy it during the job.

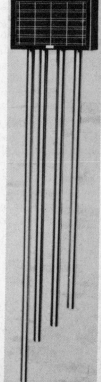

A modern grille pattern distinguishes this chime with two notes for front door, one note for rear.

Surface-mounted model has electronic mechanism with volume control installed in the cabinet. It is adjustable for eight or four notes for front door, one for rear, different note for third door. Transformer is included.

Nu Tone, Inc.

This chime comes with a transformer, has eight or four notes for front door, one note for back door, and separate tone for third door.

CHAPTER FIFTEEN

FUSES

A FUSE is simply a short piece of metal wire or ribbon that will melt when a predetermined amount of electric current flows through it. It is enclosed in a shell to prevent the melted wire from dropping free or spattering.

The type you are most likely to handle is the *plug fuse*. This form is made in ampere ratings of 30 and less for use in house wiring circuits. It screws into a socket like a light bulb (but never screw one into a bulb socket) and has a transparent window to reveal whether or not it has "blown." The melted piece of wire may be visible, plainly parted, or if a heavy short caused the fuse to blow, the window may be fogged or discolored by the metal vapor. In either event, the blown fuse looks distinctly different from the others.

As plug fuses are made in several different amperage ratings, it is possible but *definitely dangerous* to screw a high-amperage fuse into the socket formerly occupied by a lower-amperage one. The higher-rated fuse would then withstand a greater current load without blowing, but the wiring would not. The result: heated wiring within the house walls and a very serious fire hazard. To prevent this, the nontamperable fuse was developed. An adapter supplied for the fuse screws into the socket and cannot be removed. The fuse, often called a Fustat, screws into the adapter. This way, different size sockets are provided for the different ampere ratings, and the wrong amperage fuse

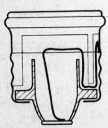 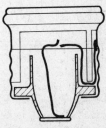

Plug fuse, which is screwed into socket in fuse box, contains a thin strip of alloy that softens and breaks when there is an overload of current. Drawing at left shows strip in working fuse; at right, after it has been broken by overload.

cannot be used in any socket. This is a potent safety factor and a convenience as well, as it automatically prevents errors.

The cartridge fuse is cylindrical in form with metal caps or knife-blade terminals at each end. Those rated at 60 amperes or less are of the cap or "ferrule" type. The round metal ends snap into spring-clip contacts. Above 60-ampere rating, the tips are of the knife-blade type, made of flat metal ex-

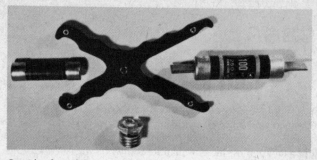

Cartridge fuses (left and right) are used for high amperage. Fuse at right, for highest amperage, has flat contact blades at ends. Small plug fuse, most common type, is shown at bottom center. Plier-like device is fuse-puller for extracting cartridge fuse from clips when it has blown.

tending from the round metal caps. The blades, too, fit into spring-clip contacts.

The time-lag fuse is similar to an ordinary fuse except that it will withstand an overload for a few seconds before blowing. This is a very handy type to use with power tools and motor-driven appliances. The reason: an electric motor that draws as little as 6 amperes while running may use as much as 30 amperes for a few seconds in starting. Thus, on a 20-ampere circuit it may blow a fuse before it has a chance to reach normal running condition if ordinary fuses are used. The time-lag fuse solves the problem without endangering the house wiring. The wiring, like the time-lag fuse, can withstand a moderate overload for a few seconds with no danger of overheating.

The circuit breaker does the same job as a fuse, but is actually an automatic switch that opens and shuts off the current when an overload occurs. It is not replaced, but simply reset, much as a toggle switch is operated. Because of its convenience, the circuit-breaker panel is becoming a popular substitute for the old familiar fuse panel.

What to Do When a Fuse Blows. *Do not* rush to replace the fuse or reset the circuit breaker. Try to find out first what caused the fuse to blow. If you don't remove the cause, the new fuse will very likely blow as soon as you insert it.

If the fuse blew out just as you plugged in or turned on an appliance, that appliance may have increased the total load on the circuit enough to exceed the capacity of the fuse. Or, the appliance may have had a short in it. In either event, disconnect it before replacing the fuse. (Procedures applied to blown fuses also apply to circuit breakers.)

If the lights are out in the fuse-panel area, take a flashlight with you when you go to replace the fuse. If the fuse panel is in the basement, remember that a damp floor can be a dangerously efficient electrical conductor. Wear rubbers and stand on a dry board if there is any sign of this type of hazard. Gloves are also a wise precaution. While modern fuse panels are designed to minimize the danger involved in changing a fuse, many older types contain an array of exposed "hot" terminals with the attendant risk of accidental contact. And there's nothing wrong in taking precautions anyway.

If there is a separate main switch, shut it off after you spot the blown fuse, then replace the fuse and turn on the current again. Many modern panels carry the fuses in removable "blocks." When the block carrying the blown fuse is pulled out with the handle provided for the purpose, it acts as a switch to cut off current to the circuits its fuses supply. The fuse may then be replaced in the block while the block is still removed from the panel. When the block is pushed back in place, it automatically switches the current on again.

In all this procedure use just one hand. Keep the other hand free, not touching the wall or anything else. Or keep it in your pocket. The risk in using both hands: if you accidentally touch a hot wire with each hand at the same time,

Most fuse cabinets and switch units generally follow this style. The cover, which contains a snap-shut door, can be removed by taking out four or more screws. This exposes connection terminals inside box for wiring work.

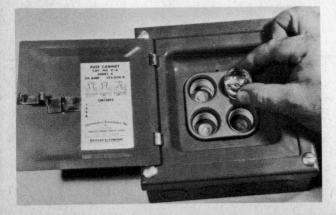

When cover is in place, all terminals are concealed, and you can screw fuses in or out without fear of contact with terminal screws. Wiring diagram on inside of door is common on most fuse and circuit-breaker units.

and if those wires happen to be on opposite sides of the circuit, as much as 240 volts may pass through your body—instead of through one hand. Either possibility should be avoided.

General Tips. Mark the fuse panel to show what amperage fuse belongs in the different sockets. An easy way to do this is to type or write in ink the fuse ratings on small paper squares and rubber-cement them to the fuse-panel diagram. If there is no diagram it's an easy matter to make one with pencil and ruler. It should show the fuses as crosses or circles, viewed as you see them from the front of the panel.

RECOMMENDED FUSE SIZES FOR HEAVY APPLIANCES

TYPE OF APPLIANCE	TYPICAL WATTS	USUAL * VOLTAGE	SIZE WIRES	FUSE SIZE RECOMMENDED
Electric Range	12,000	115/230	3 No. 6	50-60 Amp.
Dishwasher	1200	115	2 No. 12	20 Amp.
Garbage Disposer	300	115	2 No. 12	20 Amp.
Refrigerator	300	115	2 No. 12	20 Amp.
Home Freezer	350	115	2 No. 12	20 Amp.
Automatic Washer	700	115	2 No. 12	20 Amp.
Automatic Dryer	5000	115/230	3 No. 10	30 Amp.
Rotary Ironer	1650	115	2 No. 12	20 Amp.
Water Heater	Check with utility company			
Power Workshop	1500	115	2 No. 12	20 Amp.
Television	300	115	2 No. 12	20 Amp.
20,000 Btu Air Conditioner	1200	115	2 No. 12	20 Amp.
Heating Plant	600	115	2 No. 12	15-20 Amp.
Central Air-Conditioning System	Check with utility company			
Space Heating	Check with utility company			

It's a very helpful idea, too, to write in the room or rooms supplied by the fuses on the diagram. If you don't know which rooms they supply, you can find out by removing one fuse at a time (using all precautions noted earlier) while a helper checks the house and tells you the areas affected.

Always keep a supply of fuses on a shelf or in a cabinet close to the fuse panels. And be sure there are at least two of each amperage rating used in the panel. This reduces the temptation to substitute one of the wrong rating.

While there are test-light methods of locating a blown fuse when for any reason it cannot be spotted visually, these carry added shock risk for the inexperienced. If you can't locate the blown fuse in a small fuse panel, the simplest thing to do is replace all the fuses (following the precautions previously noted) and bring the removed fuses back with you to a brightly lighted area.

There you can almost always spot the blown one by eye. If you can't there's a simple way to use the battery-and-bulb test rig described under Testing Wiring. With one wire of the rig attached to a battery terminal (and leading to the flashlight bulb socket), hold the other wire from the socket against the threaded shell of each fuse and touch the bottom contact of the fuse to the other battery terminal. If the bulb lights, the fuse is good. If the bulb fails to light, the fuse is blown. The removed fuses, minus the blown one, can go back to their shelf near the fuse panel. And one final tip: don't leave any blown fuses among the good ones.

CHAPTER SIXTEEN

WORKING WITH ELECTRIC MOTORS

You DON'T have to be an expert electrician to troubleshoot the electric motors about your house. Once you learn the language of motors, and then follow through with an orderly check, you can spot the trouble and, nine times out of ten, correct it with simple tools and materials that are no farther away than your workbench.

This chapter will tell you the first things to look for if a motor runs hot, loses power, or plays dead. It will give you tips for making simple repairs. The information applies to all house motors, from the midgets in the razors and cake mixers to husky ones in oil burners, major appliances, and shop tools.

If a motor won't start, check the socket connection with a lamp—just to be sure it's not the motor's fault. If there is juice at the plug, check whether it is reaching the motor itself by rigging a simple test lamp.

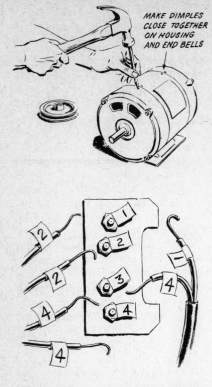

MAKE DIMPLES
CLOSE TOGETHER
ON HOUSING
AND END BELLS

Mark the body of the motor housing and the end bells before removing them. When reassembling the motor, you will be able to line up the punch marks. Then disconnect the insulated terminal plate and index the wires and terminals with dabs of matching paint or numbered strips of tape.

Motor Won't Start, No Sound. First, check to be sure there is juice at the plug. This seems obvious but, like being out of gas in a car, is the commonest cause of trouble. Pull the motor cord out of its receptacle and plug in a lamp there. If it doesn't light, check your fuse box.

Often the difficulty is that current is not reaching the motor itself. To find out, make a simple test lamp—a light bulb in a socket with a section of lamp cord attached. Separate the two wires at the free end of the cord to give you a pair of convenient probes. Bare just enough wire at their tips to make contacts.

If there's a switch on the motor circuit, turn it on. Now touch the motor connections or *terminals* (one probe tip on each). If the bulb doesn't light, go back to the switch and check it out. If the trouble is a defective toggle switch, it's usually safer and simpler to replace it. If no juice gets through to the switch, go back and check the cord plug connections. When the trouble isn't there, replace the cord. If the bulb lights at the motor terminals, it's time for internal testing.

Go slow when you remove the end plates or *bells* of a motor that has a housing. First, remove any shaft fittings (pulleys, collars or sleeves). Then mark the body of the housing and the bells by making punch marks close together. This will help you reassemble them in the same position.

Disconnect the insulated terminal plate, if there's one attached to a bell, and push it inside the housing. Index the wires and terminals with dabs of matching paint or similarly numbered strips of tape. Again, this insures correct reassembly. Slip off a bell, the one with the terminal box where there is one. If, after removing the bolts, you find that it's still wedged in the body of the housing, use a small cold chisel and a hammer—not a screwdriver—to work it free. Too much leverage at one point may damage the bearings, so tap gently all around the rim. Note the location of washers when you pull the core, or *rotor*.

Are brushes passing current? Removing the bell may expose a pair of *brushes* and a *commutator*. Check the wires or *pigtails* leading to the brush holders. One of them may have worked loose from its connections or worn bare. A brush may be sticking in its holder, or a weak spring failing to press it against the commutator. Recouple loose brush wires or pigtails. Wrap several layers of tape over any bare spots that have shorted or grounded out. Clean a binding brush with carbon tetrachloride. Replace both holder springs if one is weak—they are usually set loosely in holder pockets, behind plunger-like brushes. Again, they may bear down on the outer surfaces of lever-like holders to which the brushes are rigidly attached. Take an old spring along as a sample if you shop for a new pair at a motor-supply house. Or specify the motor or appliance model when you order from the manufacturer. Don't use springs of a different type, because too much or too little tension may cause excessive wear or poor contacts. If brushes are badly worn, replace them.

Check the commutator. If the brushes are okay, look for a loose connection where the rotor windings are soldered to the commutator segments. If the motor has stopped with one of its brushes contacting such a segment, it will refuse to start again. In other positions it will flip over, but with poor power and excessive sparking.

If you find a loose connection and are handy with a soldering iron, resolder the wire in the slot or to the stud at the rotor end of the corresponding commutator segment. Take care not to touch the coils with the hot iron. Then

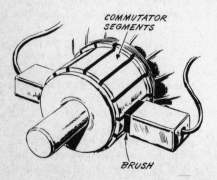

COMMUTATOR
SEGMENTS

BRUSH

Check the brushes and commutator, which will be exposed after the bell has been removed. Be sure the wires leading to the brush holders are well connected and brushes contact the commutator.

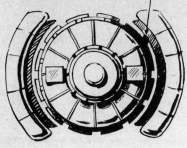

FOREIGN MATTER MAY LOCK ROTOR

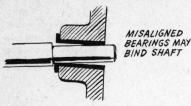

MISALIGNED
BEARINGS MAY
BIND SHAFT

If the motor hums but won't run, check
for foreign matter between the rotor and
stator. See if the shaft turns freely; mis-
aligned bearings may be binding it.

press any slack against the body of the coil and brush on enough shellac to
hold it in place. Don't run the motor until the shellac has dried.

If you don't want to tackle the soldering yourself, take the rotor to a motor-
repair shop.

How about stator connections? Inspect the wires leading to the winding of
the motor's outer, stationary magnet, or *stator*. If one of them has come off
a terminal post, fasten it back in place. When a wire is broken, bare the ends,
twist them together and solder the splice. Then wrap with tape. If an exposed section of wire is producing a short, or grounding through the motor
housing, cover it with tape. Where it's hard to wind full-width tape around
fine wire, slit off slim strips and you'll have little trouble.

One type of motor without brushes or commutator will also play dead
silently. This is the shaded-pole design easily recognized by two heavy loops
of bare copper wire fitted in slots on the stator poles. The stator-connection
check just described is the only one you need make with these motors; they
have only two wires running from their terminal posts.

If the above checks fail, there's an internal break or burn-out in one of the
motor windings. Better let a pro take over.

Motor Won't Start, but it Hums. Something inside may be causing a locking
action—a loose brush or blower fan, or dirt or grit that's worked its way between a closely spaced rotor and stator. Possibly the bearings have shifted
and are badly out of alignment. Again, an attached appliance or tool may
have a broken and wedged part, or seized bearings.

Important: Shut off the current quickly if a motor hums but doesn't run.
Then try turning the shaft by hand to make sure there is no mechanical inter-
ference or drag.

A drop in line voltage, a very long extension cord, or one of average length

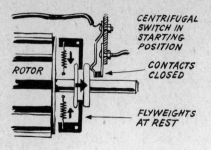

CENTRIFUGAL
SWITCH IN
STARTING
POSITION

CONTACTS
CLOSED

ROTOR

FLYWEIGHTS
AT REST

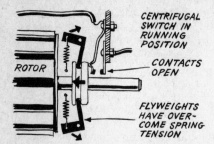

CENTRIFUGAL
SWITCH IN
RUNNING
POSITION

CONTACTS
OPEN

ROTOR

FLYWEIGHTS
HAVE OVER-
COME SPRING
TENSION

Centrifugal switch looks like this in start-ing and running positions. Contacts must be able to close and open properly, so check for gum on the shaft or washers that project too far.

with skimpy wires, may not deliver enough electricity to run the motor prop-erly. Look, too, for poor terminal connections.

If there's adequate voltage and the shaft turns freely, a low hum tells you that current is getting through some windings but not others. Give motors with brushes and commutators the same internal checks you would the silent kind.

If a motor has no brushes and commutator, you'll find four wires taking off from its terminals. (One exception is the shaded-pole type already de-scribed.) One pair leads to the main, or *running* coils of the stator. The other two are connected, by way of a centrifugal switch on the major shaft, to a sec-ond set of stator windings called *starting* coils. Check the switch; it looks much like any other governor. In the starting position, juice is directed through the starting windings. Then, as the motor revs up, a pair of weights fly out, compressing a spring and breaking the contacts. Gum on the shaft, or washers that project too far, may be holding these contacts apart at all times. Clean the shaft with kerosene applied with a brush; then oil sparingly. Or remove a washer that prevents contact. Sandpaper contacts lightly if they're dirty.

Motor Overheats, or Blows Fuses. Most motors need a flow of air through their housings to keep them cool. In one that has netting or perforated sheets behind the air vents in its end plates, make sure this screening isn't clogged with dirt and dust.

Perhaps bearings are dry or improperly aligned, or a rotor is dragging. Stiff bearings impose an unnecessary load on a motor. Loosen the bells, and try shifting them. Where a flange on the housing prevents this, try snugging up the bolts. If one or more of them is loose, the bearings may be in a plane not perpendicular to the shaft. A dragging rotor may be due either to mis-aligned or worn bearings. In the latter case, you can have them replaced or rebushed if the motor is worth it.

Is enough voltage getting to the motor terminals? A motor that isn't fed enough voltage may draw more current than it can safely handle. Assuming it's on a properly fused circuit, the fuse will blow. If the fuse has too high a rating, the motor will burn out.

Likewise, conditions that block a flow of current to or through a coil may not stop a motor, but other windings will work harder than they should. Eventually, they will destroy themselves. You've already learned what checks to make to spot a dead winding.

Motor Squeaks, Rumbles, or Clicks. Look for stiff bearings, a major cause of motor squeaks. A motor that rumbles either has worn bearings or an out-of-balance rotor.

Clicks, except for the reassuring kind that tell you that a starting switch has just broken contacts, indicate brush or commutator troubles. They will be accompanied by excessive sparking. Turn the shaft by hand. Possibly the brushes are stubbing their toes against high mica insulation between the commutator segments. Again, the soft brush material may have worn down to a point where the hard brush-holder is dragging against the drum. Brushes should be replaced long before they are reduced to nubs. Apart from the risk of holders dragging against the commutator and scoring it, spring tension may ease off enough to cause arcing and loss of power.

If brushes don't seat against segments correctly, or if you are replacing them, take a strip of fine sandpaper—not emery cloth—and wrap it about the commutator, abrasive side out. Then rotate the shaft by hand. At the same time, press lightly on the brushes. This will produce matching concavities on the brushes. Wipe away any trace of grit before you reassemble the motor. When doing this, make sure the holders are properly aligned.

If mica has crept up between the commutator segments and is striking the brushes, break off the end of a hacksaw blade and grind away its tooth set. Pull the rotor assembly out of the motor and, clamping it in a vise with protective pads of wood or leather, resaw the slots between segments.

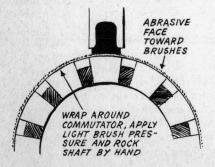

ABRASIVE
FACE
TOWARD
BRUSHES

WRAP AROUND
COMMUTATOR, APPLY
LIGHT BRUSH PRES-
SURE AND ROCK
SHAFT BY HAND

Brushes should be seated correctly against segments of the commutator. If they are not, wrap a strip of fine sand-paper around the commutator, abrasive side out, and rotate the shaft by hand, applying light pressure on the brushes. This will produce matching concavities on the brushes.

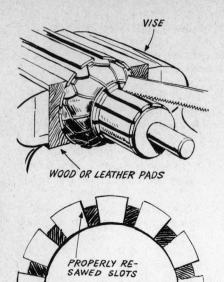

VISE

WOOD OR LEATHER PADS

PROPERLY RE-
SAWED SLOTS

Mica may creep up between the commutator segments and strike the brushes.
Clamp rotor assembly in vise and resaw
slots between segments with a hacksaw
that has had its tooth set ground away.

Care of Electric Motors. Overenthusiastic oiling and greasing probably knock out more home motors than abuse. Working its way into commutators, brushes and other contact surfaces, the excess can attract dirt and dust, bonding the particles together to form an insulative film that puts the contacts out of commission. Lubricate all motor bearings sparingly. A few drops of SAE 20 oil for 1/6 hp. up, or the water-thin variety, packaged in squirt cans for the midgets, will keep bearings running free for months. So will a single clockwise twist of a grease-cup cap. Leave the lubricated-for-life type strictly alone.

One more tip. Bearing oil cups often have filtering wicks. Occasionally one of these may clog. When oil backs up behind a wick, remove the felt or other filtering materials carefully with tweezers, and wash it in kerosene. Then let it dry before putting it back in the cup and replenishing the supply of lubricant.

Vacuum out the fluff. You've seen how dirt and dust may stall a motor, cause overheating or build up an insulative film. They can do it, even without help from excess oil. This kind of fluff also presents a fire hazard. Dirt will cause excessive brush and commutator wear. Keep your motors clean. Vacuum the larger ones, and use a soft brush to dust off the small ones. Don't neglect the midgets tucked under phonograph turntables, in the housings of slide projectors and other hard-to-reach spots. It may be convenient to forget them, but house dust doesn't and is more than a match for their flea-power.

Extension cords. A further word of caution about these. Never put a motor farther than 200 feet from a power outlet. And, finally, remember that good insulation pays off. Use SJ type cord for moderate duty, and type S for heavy service or where the cord is subject to oil or abrasion.

INDEX

A

adapter, 44, 79, 127
Adequate Wiring Bureau, 12
air conditioning, 47, 49, 51, 53, 83
Alabax, shower light, 49
alternating current, 7–8
alternators, 8
American Wire Gauge, 21
amperages, 127–128
amperes, 7, 8, 50
appliance circuits, 13
appliance cord, 81
appliance unit, 75
armatures, 8, 101, 102
armored cable, 17, 18, 19, 20, 23, 24, 33, 35, 36, 39, 41, 42, 55, 56, 57, 70, 72, 74
awl, 16

B

back-wired receptacles, 44, 45, 46
bathroom wiring, 49
battery-and-bulb testing rig, 116, 131
bedroom lighting, 92
bedroom wiring, 48, 105
bedside lamps, 48
bits, drill, 36
blanket, 48
boxes, wiring, 39–45
brace and bit, 17, 20, 36
brackets, 39
break-out, 52, 53
Brown and Sharp Gauge, 21
"bug light," 109
bulb plug, 79
bulb socket, 38, 90
bulbs, 7, 94
bunch splice, 26, 28
bushings, 33, 34–35, 39, 40, 41, 71
BX cable, 11, 19, 22, 23, 24, 33, 34, 35, 36, 37, 58, 64, 72, 119

C

cable, 9, 10, 11, 17, 18, 19, 20, 21, 22–24, 33, 34, 35, 36–37, 38, 39–40, 41, 42, 43, 50, 51, 52, 55, 56, 60, 70–71, 72, 92, 99, 114–115, 119

cable cutting, 33, 34
cable designation, 24
cable ripper, 42
cable stringing, 70–71
cable support, 72–73
cable, waterproof, 23
ceiling boxes, 88, 89
ceiling fixtures, 87–94
ceiling lights, 10, 38, 46, 48, 49
ceiling tracks, 91
cellar steps wiring, 49
chandeliers, 91, 93
chimes, 4-note, 124
chimes, surface-mounted, 126
chimes, 2-chord, 126
chimes, wiring, 124
chimes with transformers, 126
chisel, 20, 134
circuit breakers, 9, 22, 24, 79, 91, 128
circuit tester, 19, 116–117, 125
circuit tracing, 55
circuits, 9, 10, 12, 13, 21, 22, 32, 34, 41, 48, 50, 51, 52, 53, 54–55, 58, 61, 73, 74, 77, 91, 128
circular mil, 21
clamps, 38, 39, 43
clock-hanger outlet, 47, 75
clock radio, 48
clock, wall, 47
closet light, 48, 85
Code book, 14–15
coils, 8, 123, 136
coils, motor-running, 136
coils, motor-starting, 136
color coding, 9–10, 54
conductors, 22, 32
conduit, 9, 16, 17, 18, 20, 23, 37, 44, 72, 74
conduit adapter, 44
conduit bender, 16, 23, 37, 45
connecting wires to service panel, 35, 40
connectors, solderless, 20, 23, 28–29, 34, 35, 39, 40, 41, 42, 43, 44, 71, 89, 121
contact, 68, 136
cooking appliances, 48, 107, 109–110
cord, asbestos-insulated, 83
cord attachment to terminal screw plug, 80

cord, five types of, 83
cord plug, lever-operated, 79
cord repairing, 84
cord-selection chart, 86
cord splices, 84
cord switch, 100
cords, 21–22, 81, 82–86
core, transformer, 8
coulombs, 7
couplings, 23, 37, 44
Cresflex nonmetallic cable, 23
current, 7–8, 10, 18, 19, 30, 54, 56, 60, 68, 84, 90, 91, 116
current, shutting off, 68

D

Despard, Victor, 52
dimmer control, 92, 93, 105–106
dining-room wiring, 47–48
direct current, 7
distribution panel, 22
door buzzers, 7, 123–127
door chimes, 123–127
door switch, automatic, 48
doorbell-buzzer wiring, 124–126
doorbells, 59, 62, 105, 123–127
drill attachment, right-angle, 36
drills, 17, 20, 36
drip loop, 32
duplex connector, 41
duplex outlet receptacle, 52–53, 57, 61, 78
dust-and-moisture-proof receptacle, 76

E

electromagnets, 8
electrons, 7
entrance head, 12, 31, 32
entrance panel, 12, 18, 31, 73
entrance switch, 31
entrance wires, 30, 31
extension cords, 84–85, 138

F

family room wiring, 48–49, 53
fans, 53, 109
faucet, outdoor, 113, 114
feed-in, 58
file, 18
"fish tape," 16, 70, 71, 73
fishing wire through holes, 73
fixture boxes, 10, 37, 39, 40, 42, 43, 87
fixture repairs, 90
fixture studs, 87

fixture wiring, 90
fixtures, new work, 87
fixtures, old work, 87
flexible armored cable, 23
flexible cord, prohibitions against, 83–84
Flexsteel, 23
floor outlet, 76
floor polisher, 48, 49
fluorescent light ballast, 118–119
fluorescent light channel box, 118–120
fluorescent light installation, 119–120
fluorescent light starter, 118, 119, 120
fluorescent light tubes 118–119
fluorescent lighting fixtures, 62, 88, 91, 118–122
fountains, 109
flux, 26
4-light ceiling fixture, 94
4-plug outlets, 75
4-way switches, 98–99
4-wire cable, 99
friction tape, 20, 27, 28, 40, 71
fuse adapter, 127
fuse block, 129
fuse blow-out, 128–129, 136
fuse box, 30, 31, 54–55, 56, 90, 127, 129, 130
fuse-box diagram, 130
fuse, cartridge, 128
fuse, plug, 127–128
fuse supply, 130
fuse, time-lag, 128
fuse tips, ferrule, 128
fuse tips, knife-blade, 128
fuses, 9, 22, 24, 35, 55, 68, 73, 74, 127–131
fuses, tips on, 130–131
Fustat, 127

G

game-area lighting, 110
gang box, 38
garden lighting, 111, 112, 113
grommet, insulated, 120
ground terminal, 51
ground wire, 10, 11, 22, 30, 31, 42, 51, 52, 54, 77, 91
grounding, 10, 30–31, 47, 50, 56, 74–77, 81, 117
grounding testing, 117

H

hacksaw, 17, 18, 20, 23
hair-dryer, 48, 49

hallway wiring, 49
hammer, 20, 35, 70, 71, 119, 134
hangers, 47, 75, 87, 89
Harvey Hubbell, Inc., 81
heater cord, 22, 83
heaters, 7, 22, 82, 110
heating cable, 114
heating pad, 48
heating tape, 113–114
heavy-duty cord, 22, 58, 82, 83
heavy-duty plug, 82
heavy-load wiring, 50–53
heavy-wire connector, 29
hedge trimmers, 81
hi-fi, 46, 48
Hi-Lo switch, 104
hickey, 88, 90
Hold-It, 87
hot plates, 22
"hot" wires, 10, 11, 22, 30–32, 33, 51, 52, 53, 129
hotbeds, 115
house wiring, 9–10, 30–49, 50–53, 54–56

I

ice-dams, 113–114
indirect lighting, 46, 48
inspection, building department, 30, 32, 74
insulating tape, 20, 27–28
insulation, 19, 20, 21–22, 23, 24, 25, 26, 28, 30, 34, 35, 40, 41, 44, 54–55, 81
insulation stripping tools, 19
insulators, 32
interchangeable devices, 45, 47, 51, 52, 53
irons, 13, 22, 82

J

joists, 36, 37, 39, 43
junction box, 37–39, 56, 123

K

kitchen wiring, 48, 51
"knob and tube wiring," 54, 55
knockouts, 18, 20, 35, 38, 71, 73, 119

L

lamp cord, 21–22, 82, 83, 100
lamp socket, 85
laundries, automatic, 13, 49, 50
laundry/utility room wiring, 49
lawn mowers, 81
light-control switch, 104, 105

lighting, 13, 91–94, 107–108
lighting circuits, 13
lighting, dramatic, 108–109
lighting, fall & winter, 109–110
lighting outlets, 13
lighting, science of, 91–94
lightning, 10
Lightolier, 91
link surface wiring, 66
living-room wiring, 46, 51, 53
lock switch, 49
locking plate, 52
locknuts, 35, 40, 71
loops, wire, 40, 41
lugs, power take-off, 31
lumens, 94
Luxtrol switch, 104, 105–106

M

magnet, horseshoe, 8
magnet wire, 8
magnetism, 8
main disconnect, 31
main switch, 9, 12, 51, 54, 68, 90
measuring electricity, 7
meter, 9, 11, 12
meter socket, 9, 11
mixer, 132
motor bells, 133–134
motor brushes, 134, 135, 136
motor, care of, 138
motor commutator, 134–135, 136, 137
motor contacts, 136
motor, non-starting, humming, 135–136
motor, non-starting, non-sounding, 133–135
motor, overheated, 136–137
motor pigtails, 134
motor, shaded-pole, 135, 136
motor, squeaking, rumbling, clicking, 137–138
motor switch, centrifugal, 136
motor terminals, 133
motor-vacuuming, 138
motors, 78, 132–138
mounting arrangements, 39
mounting new fixtures in old box, 89–90
movie projector, 46, 48
multi-outlet plug, 79
multi-outlet flexible strip, 58, 59–60, 61, 79
multi-outlet plug-in strips, 58, 59–60, 61, 79

multi-outlet raceway, metallic, 58–59, 60
multi-outlet wiring system, 60–63, 119

N

National Board of Fire Underwriters, 14
National Electrical Code, 11, 12, 14–15, 23, 30, 32, 48, 56, 60, 74, 83, 84
National Fire Protection Association, 14
neutral strip, 30
neutral wire, 10, 22, 30, 31, 33, 51, 52
new wiring, 37–45, 87
new wiring, carpentry & mechanics of, 37
night light, 45, 48, 49, 52, 79
nite-lite adaptor plug, 79
nite-lite plug-in, 79
nonmetallic sheathed cable, 23, 24, 37, 42, 43, 55, 56, 72, 74, 77
nonmetallic surface extension, 58, 59, 67
nonmetallic surface wiring system, 64–67
nursery wiring, 48, 105

O

oil burner, 5, 132
oil-resistant cord, 83
old wiring, checking, 54–55, 56
old-work hanger, 89
outdoor lighting, 107, 108, 110
outlet-box cover plates, 38, 54, 64, 65
outlet box, rectangular, 37, 39
outlet boxes, 37–45, 52, 54, 87–91, 123
outlet connections, 71–72
outlet, cutting hole for, 71
outlet repairs, 68–69
outlet safety, 76
outlet, switch, pilot-light combination, 76
outlets, 10, 12–13, 16, 17, 18, 19, 20, 23, 34, 35, 37, 45, 46–49, 51, 52, 55, 56, 58–60, 63, 64, 65, 66, 68–77, 78, 85, 107
outlets, adding new, 69–77
outlets, grounding, 47, 49, 59, 63, 74–77
outlets, location of, 69–70
outlets, plug-in, 79
outlets, tamper proof, 48
outlets, weatherproof, 108
overhead wires, 30, 31, 32
overheating wires, 55
overloads, 9, 55, 73–74

P

Pass and Seymour, 48, 49
patio lighting, 111, 112, 113
permit, wiring, 30
phosphor, 118, 119
picture-window lighting, 111
pigtail splice, 24, 25
pilot light, 45, 48, 52, 76
pipe reamer, 18, 23
pipe strip, 44
plastic-covered cable, 23, 42
plastic-covered cord, 82, 83
plastic strips, 58, 59–60, 61, 64–65, 67
plastic tape, 20, 27, 28, 53
plier-cutter tools, 19
pliers, 16–18. 24, 25, 35, 41, 67
pliers, electrician's, 16, 17, 24
pliers, mechanic's, 16
pliers, small-nosed, 16–17, 41
pliers, strip-baring, 67
pliers, utility, 16
plug, circuit-breaker, 79, 81
plug, quick-connect, 79
plug, round with grip-neck, 79
plug, self-connecting, 120
plug, series-wired, 79, 82
plugs, 19, 52, 68, 74–75, 78–82, 120
plugs, grounding, 74–75, 81
plugs, molded, 78, 81
plugs, slim, 78
plugs, terminal screw, 78–81
pocketknife, 16, 19, 40, 43, 44, 81
"potential," 19
power company, 7, 9, 11, 30, 50, 51
primary coil, 8, 123
pull-chain switch, 38, 39, 48, 100, 119
punch blade, 16
push-in connection, 44–45, 47

R

raceway, 9, 10, 17, 18, 20, 39, 49, 55, 58–67, 78, 79, 119
raceway, baseboard, 61
raceway channel, 59, 60, 61, 62, 66
raceway, single-current, 61
radios, 13, 46, 48, 68, 82
ranges, 13, 50
razor, 48, 49, 132
relay unit, 103
rocker switches, 46, 49
Romex nonmetallic cable, 22, 23, 24, 37
room wiring and lighting tips, 45–49
rotor, 134–135

rubber-covered cord, 82, 83
rubber tape, 20, 27, 28
rule, measuring, 18

S

safety codes, 11, 12, 14–15, 23, 30, 32, 48, 50, 51, 58, 59, 106, 107, 108, 124
saw, compass, 17
saw, keyhole, 20
screwdrivers, 16, 18, 35, 38, 43, 52, 70, 71, 90, 119, 130
screwdrivers, electrician's, 16, 18
screw-on connector, 28–29
screws, 9, 10, 17, 41, 51, 52, 69, 75, 77
screws, brass, 10, 41, 51, 52, 75
screws, chrome, 10, 41, 51, 52, 75
screws, green terminal, 75, 77
screws, tightening, 68
secondary coil, 8, 123
service entrance, 9, 12, 21, 22, 30, 50
service head, 9
service installation, 30–36
service panel, 9, 10, 12, 30, 33, 34–35, 36, 37, 40, 50, 73
setscrew connector, 29, 71
sewing machine, 49
shears, aviation, 18
sheet metal, 18
short circuits, 9, 54, 91, 127
shower light, 49
Simonds mower blade file, 18
skating rink, 113
Smith-Gates Corp., Farmington, Conn., 115
snaking, 55, 57
snap-on strip, 61, 62
snips, metal, 18–19
solder, 19, 25, 26–27, 29
solder, rosin-core, 26
soldering, 19, 24, 25, 26–27, 41, 90–91, 134
soldering gun, 19
soldering iron, 19, 26, 90–91, 134–135
soldering splices, 26–27, 28, 87
soldering tools, 19
sole plate, 37
spacer strips, 58, 60
splice, exposed cable, 56
splices, 10, 19, 20, 24, 25–29, 38, 39, 44, 56, 62, 84
splicing, 17, 24–29, 40, 42, 89
split circuits, 51, 52, 53
split receptacles, 51
spotlight, 91

Stanley's aviation snips, 19
staples, 20, 36, 41, 42, 43, 72, 83, 123
stator, 135
sterilizer, 48
stranded wire, 21
straps, 37, 41, 42, 43, 44, 72, 87
strip gauge, 45
strip-lighting, 93
strips, 45, 58, 59–60, 61, 64–65, 67, 79, 93
studs, 37, 87, 88
sun lamp, 49
Superior Electric Co., Bristol, Conn., 115
surface cooking burner group, 50
surface-mounted box, 38
surface-mounted ceiling lights, 93
surface wiring, 23, 55, 57–67
surface-wiring installation, 59–67
surface wiring, sectional, 66
switch box, 17, 20, 38, 39, 76
switch buttons, luminous, 49
switch installation, 95–97
switch, master selector, 103
switch, mercury, 46, 102, 105
switch, photocell, 101
switch, press, 101
switch, remote-control, 103
switch replacement, 99–100
switch, single-pole, 105
switch, 3-way, 91, 97–98, 99, 105
switch, time-clock, 101, 108
switch, time-delay, 100
switch, toggle, 102, 128, 133
switches, 10, 12, 16, 18, 24, 31, 35, 37, 38, 39, 41, 43, 45–49, 52, 90, 91, 95–100
switches, low-voltage, 103, 105
switches, special purpose, 100–106
switches, weatherproof, 108

T

tap splice, 15, 26, 27, 84
taping, 27–28, 41, 42
test light, 18, 19, 90, 117, 133
thin-wall conduit, 23, 44, 74
3-opening Despard, 52
3-pronged grounded plug, 81
3-prong plug receptacle, 74, 77
3-wire adapter, 79
3-wire cable, 22, 51, 52, 53, 77
3-wire circuits, 30, 32, 41, 51
3-wire connections, 32, 33
3-wire grounding outlets, 47, 49, 59, 63
3-wire plug, 79

3-wire power-tool cord, 83
3-wire raceway, 61
toaster, 7, 13, 22
toggle, 49
tools, 13, 16–20, 62, 81, 83, 132
transformer, 7–8, 123, 124–125
transmission, 7–8
TV sets, 13, 48, 53
Twist-Lock plugs, 81
2-prong plugs, 75, 77
2-way cable, 22, 42, 52, 95
2-wire entrance, 73
2-wire outlet, 63
2-wire plugs, 79
2-wire power-tool cord, 83
2-wire raceway, 62
Type AC cable, 24
Type HPD Underwriters' cord, 83
Type NM cable, 24
Type NMC cable, 24
Type S Underwriters' cord, 83
Type SJ Underwriters' cord, 83
Type SJO Underwriters' cord, 83
Type SO Underwriters' cord, 83
Type SP Underwriters' cord, 83
Type SPT Underwriters' cord, 82
Type TW color-coded wire, 44
Type UF cable, 24
Type UF-NMC cable, 24
Type USE cable, 24

U

underground cable, 22, 23, 24
underground fused cable, 24
underground service-entrance cable, 24
Underwriters' cord, 82–83
underwriter's knot, 79, 80
Underwriters' Laboratories, Inc., 10, 23, 24, 58
utility room wiring, 49

V

vacuum cleaners, 13, 46, 48, 49

vaporizer, 48
vibrator, 49
voltages, 32, 34, 50–51, 100, 123, 136, 138
volts, 7, 8, 50

W

wall fixtures, 10, 46, 47, 87–94
wall fixtures, recessed, 92
wall oven, 50
water heaters, 50, 51
watts, 7, 50
Western Union splice, 25–26
wire, aluminum, 21
wire baring, 81
wire, black, 10, 24, 30, 31, 33, 35, 41, 51, 52, 53, 54, 90, 91, 92, 123
wire connections, 40–41, 42, 44
wire, copper, 21, 22
wire, red, 30, 31, 33, 51
wire sizes, 21, 50, 55, 58, 73, 75, 82, 123
wire, white, 10, 24, 35, 41, 50, 51, 52, 53, 54, 90, 91, 92, 123
Wiremold, 58, 62, 119
wiring, ceiling-to-wall, 72
wiring devices, location of, 46
wiring errors, 54
wiring, floor-to-floor, 37
wiring, house, 9–10, 30–49
wiring, planning, 12–13
wiring, outdoor, 107–115
wiring safety, 9, 10–11, 14–15
wiring techniques, 24–29
wiring, testing, 116–117, 125–126
wiring, 240-volt, 50
wiring, underground, 107
wiring, wall-to-ceiling, 72
woodworking tools, 20
workshop wiring, 49, 51

Y

Yankee push drill, 17